Uses and Limitations of Observations, Data, Forecasts, and Other Projections in Decision Support for Selected Sectors and Regions

U.S. Climate Change Science Program

Synthesis and Assessment Product 5.1

August 2008

FEDERAL EXECUTIVE TEAM

Director, Climate Change Science Program ....William J. Brennan

Director, Climate Change Science Program Office ....Peter A. Schultz

Lead Agency Principal Representative to CCSP,
Associate Director for Research, Earth Science Division,
National Aeronautics and Space Administration ....Jack Kaye

Lead Agency Program Director, Associate Director for Applied Sciences,
Earth Science Division, National Aeronautics and Space Administration ....Teresa Fryberger

Product Lead, Applied Sciences Program, Earth Science Division,
National Aeronautics and Space Administration ....John A. Haynes

Synthesis and Assessment Product Coordinator,
Climate Change Science Program Office ....Fabien J.G. Laurier

NASA Point of Contact...John Haynes, NASA HQ

EDITORIAL AND PRODUCTION TEAM

Scientific Editor ....James Koziana, SAIC
Scientific Editor ....Fred Vukovich, SAIC
Scientific Editor ....Molly Macauley, RFF
Technical Editor..Christy Churchwell, SAIC
Layout/Design...Peter Gregory, SAIC

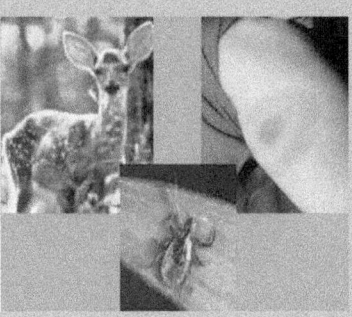

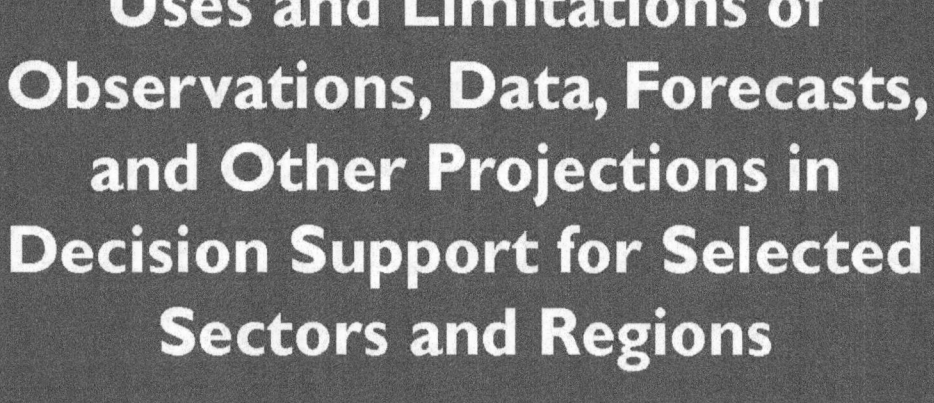

Uses and Limitations of Observations, Data, Forecasts, and Other Projections in Decision Support for Selected Sectors and Regions

Synthesis and Assessment Product 5.1
Report by the U.S. Climate Change Science Program
and the Subcommittee on Global Change Research

Lead Agency

National Aeronautics and Space Administration

Supporting Agencies

Department of Commerce/National Oceanic and Atmospheric
Administration, Department of Energy, Department of Interior/
U.S. Geological Survey, National Science Foundation, U.S. Agency for
International Development, U.S. Environmental Protection Agency

Members of Congress:

On behalf of the National Science and Technology Council, the U.S. Climate Change Science Program (CCSP) is pleased to transmit to the President and the Congress this Synthesis and Assessment Product (SAP), Uses and Limitations of Observations, Data, Forecasts, and Other Projections in Decision Support for Selected Sectors and Regions. This is part of a series of 21 SAPs produced by the CCSP aimed at providing current assessments of climate change science to inform public debate, policy, and operational decisions. These reports are also intended to help the CCSP develop future program research priorities.

The CCSP's guiding vision is to provide the Nation and the global community with the science-based knowledge needed to manage the risks and capture the opportunities associated with climate and related environmental changes. The SAPs are important steps toward achieving that vision and help to translate the CCSP's extensive observational and research database into informational tools that directly address key questions being asked of the research community.

This SAP focuses on the use of climate observations, data, forecasts, and other projections in decision support. It was developed with broad scientific input and in accordance with the Guidelines for Producing CCSP SAPs, the Federal Advisory Committee Act, the Information Quality Act (Section 515 of the Treasury and General Government Appropriations Act for Fiscal Year 2001 (Public Law 106-554), and the guidelines issued by the National Aeronautics and Space Administration pursuant to Section 515.

We commend the report's authors for both the thorough nature of their work and their adherence to an inclusive review process.

Samuel W. Bodman
Secretary of Energy
Vice Chair, Committee on
Climate Change Science
and Technology Integration

Carlos M. Gutierrez
Secretary of Commerce
Chair, Committee on Climate
Change Science and Technology
Integration

John H. Marburger III
Director, Office of Science
and Technology Policy
Executive Director, Committee
on Climate Change Science and
Technology Integration

TABLE OF CONTENTS

CHAPTER

AUTHOR TEAM FOR THIS REPORT

Executive Summary

Lead Authors: Fred Vukovich, SAIC; Molly K. Macauley, RFF

Chapter 1

Lead Author: Molly K. Macauley, RFF

Chapter 2

Lead Author: Daewon W. Byun, University of Houston

Chapter 3

Lead Author: David Renne, NREL

Chapter 4

Lead Author: Gregory Glass, Johns Hopkins School of Public Health

Chapter 5

Lead Author: Holly Hartmann, University of Arizona

ACKNOWLEDGEMENT

This report has been peer reviewed in draft form by individuals chosen for their diverse perspectives and technical expertise. This document was reviewed via an external review, Public/Federal Agency Review and a NASA Concurrence. The expert review and selection of reviewers followed the OMB's Information Quality Bulletin for Peer Review. The purpose of these independent reviews was to provide candid and critical comments that will assist the Climate Change Science Program in making this published report as sound as possible and to ensure that the report meets institutional standards.

We wish to thank the following individuals for their peer review of this report: Frank Muller-Karger (Univ. of South Florida), Roger King (Mississippi State Univ.), Chuck Hutchinson (Univ. of Arizona), Verne Kaupp (Univ. of Missouri), Richard Lawford (GEWEX Project Office), D. Lettenmaier (Univ. of Washington), Pierre-Philippe Mathieu (European Space Agency), Donald Blumenthal (Sonoma Technology, Inc.), Richard Scheffe (US EPA), Gregory Carmichael (Univ. of Iowa), Kelly Chance (Harvard-Smithsonian Center for Astrophysics), Judith Qualters (Centers for Disease Control and Prevention), Robert Venezia, (Armed Forces Medical Intelligence Center) and Jonathan Patz (Univ. of Wisconsin).

In addition, we wish to thank the following individuals for their Public/Federal Agency review of this report: Amy Kaminsky (OMB), Linda Lawson (DOT), Samuel Williamson (NOAA), Derek Parks (NOAA), Kevin Trenberth (NCAR), Thomas Armstrong (USGS), Tom Vonder Harr (Colorado State), and Marta Cehelsky (NSF)

Furthermore, we wish to thank the following individuals for their NASA Concurrence of this report: Teresa Fryberger, Jack Kaye, John Haynes, Michael Freilich, Randy Friedl, Colleen Hartman and Alan Stern

The author wishes to thank Ed Scheffner for his contribution to chapter 1.

The author wishes to thank the following for their contribution to chapter 2.
Tracey Holloway (University of Wisconsin—Madison)
Daniel J. Jacob   (Harvard University)
Jennifer A. Logan   (Harvard University)
Loretta J. Mickley (Harvard University)
Xin-Zhong Liang (University of Illinois at Urbana-Champaign)
Dr. Zong-Liang Yang   (The University of Texas at Austin)
Prof. Armistead (Ted) Russell   (Georgia Institute of Technology)
Carey Jang   (USEPA/OAQPS)
Alice Gilliland   (USEPA/ORD/NERL/AMD)

The author wishes to thank staff members at the National Renewable Energy Laboratory for their many insights and contributions to the area of renewable energy resource assessment, and to this chapter 3 in particular, including Dennis Elliott and Marc Schwartz (wind resource assessments), Ray George, Steve Wilcox, Daryl Myers and Tom Stoffel (solar resource assessments), Donna Heimiller and Anelia Milbrandt (Geographic Information Systems and model input data), and Shannon Cowlin (HOMER applications). The author also wishes to thank Peter Lilienthal (Green Island Power) who developed the HOMER model, and Paul Gilman (HOMER consultant), who reviewed that chapter and provided many useful comments.

The author would like to thank Drs. Durland Fish (Yale University) and Joseph Piesman (U.S. Centers for Disease Control and Prevention) for assistance in characterizing the development of the Lyme Disease Prevention DSS. Dr. John Brownstein was instrumental in providing access to figures in Chapter 4, showing the output of these analyses.

RECOMMENDED CITATIONS

For Chapter 1:

Kanarek, Harold. 2005. "The FAS Crop Explorer: A Web Success Story," *FAS Worldwide*, June (http://www.fas.usda.gov/info/fasworldwide/2005/06-2005/Cropexplorer.htm (accessed April 2007).

National Research Council, Board on Earth Sciences and Resources. 2007. *Contributions of Land Remote Sensing for Decisions about Food Security and Human Health: Workshop Report* (Washington, DC: National Academies Press).

Reynolds, Curt A. 2001. "CADRE Soil Moisture and Crop Models," at http://www.pecad.fas.usda.gov/cropexplorer/datasources.cfm (accessed April 2007).

United Nations Food and Agriculture Organization. No date. "Agriculture and Climate Change: FAO's Role" at http://www.fao.org/News/1997/971201-e.htm (accessed April 2007).

For Chapter 2:

Hogrefe C, LR Leung, LJ Mickley, SW Hunt, and DA Winner, 2005. "Considering Climate Change in U.S. Air Quality Management," EM: Air & Waste Management Association's magazine for environmental managers October 2005:19–23.

Jacob, D.J., and A.B. Gilliland, 2005. "Modeling the Impact of Air Pollution on Global Climate Change, Environmental Manager," pp. 24–27, October 2005, Air and Waste Management Association. Pittsburgh, PA.

Leung LR, Y Kuo, and J Tribbia. 2006. "Research Needs and Directions of Regional Climate Modeling Using WRF and CCSM," Bulletin of the American Meteorological Society 87(12):1747–1751.

LRTAP, 2007b: Task Force on Hemispheric Transport of Air Pollution, 2007 Interim Report. Available at http://www.htap.org/activities/2007_Interim_Report.htm.

Mickley, L.J., D.J. Jacob, B.D. Field, and D. Rind, 2004. "Effects of Future Climate Change on Regional Air Pollution Episodes in the United States," Geophys. Res. Let., 30, L24103, doi:10.1029/2004GL021216.

Tagaris, E., K. Manomaiphiboon, K.-J. Liao, L. R. Leung, J.-H. Woo, S. He, P. Amar, A. G. Russell, 2007. "Impacts of Global Climate Change and Emissions on Regional Ozone and Fine Particulate Matter Concentrations over the United States," J. Geophys. Res., 112 (D14), D14312.

For Chapter 3:

Lambert, Tom, Paul Gilman, Peter Lilienthal., 2006. Micropower System Modeling with HOMER. In Felix A Farret, M Godoy Simoes. *Integration of Alternative Sources of Energy*. John Wiley and Sons, Inc. Hoboken, New Jersey. 379-416

Perez, R., P. Ineichen, K. Moore, M. Kmiecik, C. Chain, R. George, and F. Vignola, 2002: A New Operational Satellite-to-Irradiance Model. *Solar Energy* 73(5), pp. 307–317.

Renné, David S., Richard Perez, Antoine Zelenka, Charles Whitlock, and Roberta DiPasquale, 1999: Use of Weather and Climate Research Satellites for Estimating Solar Resources. Chapter 5 in Advances in Solar Energy, Volume 13, Edited by D. YogiGoswami and Karl W. Boer. The American Solar Energy Society, 2400 Central Ave. Suite G1, Boulder, Colorado 80301. Pp.171–240.

The U.S. Climate Change Science Program Appendix A

Schwartz, M., R. George, and D. Elliott, 1999. The Use of Reanalysis Data for Wind Resource Assessment at the National Renewable Energy Laboratory. Proceedings, European Wind Energy Conference, Nice, France, March 1–5, 1999.

For Chapter 4:

Brownstein, J.S., T.R. Holford and D. Fish 2005a: Effect of climate change on Lyme disease risk in North America. *EcoHealth* 2:38–46.

Brownstein, J.S., D. K Skelly, T.R. Holford and D. Fish. 2005b: Forest fragmentation predicts local scale heterogeneity of Lyme disease risk. *Oecologia* 146: 469–475

Glass, G.E. 2007: Rainy with a chance of plague: forecasting disease outbreaks from satellites. *Future Virology* 2:225–229

For Chapter 5:

Carron, J., E. Zagona, and T. Fulp (2006) Modeling Uncertainty in an Object-Oriented Reservoir Operations Model. J. Irrig. and Drain. Engrg., 132(2): 104-111.

Frevert, D., T. Fulp, E. Zagona, G. Leavesley, and H. Lins (2006) Watershed and River Systems Management Program: Overview of Capabilities. J. Irrig. and Drain. Engrg. 132(2):92-97.

Grantz, K., B. Rajagopalan, E. Zagona, and M. Clark (2007) Water management applications of climate-based hydrologic forecasts: case study of the Truckee-Carson River basin, Nevada <http://cadswes.colorado.edu/PDF/RiverWare/

GrantzEtAl2006WaterManagementApps_JWRPM.pdf> . Journal of Water Resources Planning and Management.

Hartmann, H.C. (2005) Use of climate information in water resources management. In: Encyclopedia of Hydrological Sciences, M.G.

Hydrological Sciences Branch (2007) Evaluation Report for AWARDS ET Toolbox and RiverWare Decision Support Tools. NASA Goddard Space Flight Center, Greenbelt, MD, 28pp. (URL: http://wmp.gsfc.nasa.gov/projects/project_RiverWare.php)

Lettenmaier, D.P. (2003) The role of climate in water resources planning and management. In: Water: Science, Policy, and Management, R. Lawford, D. Fort, H. Hartmann, and S. Eden (Eds.), American Geophysical Union, Washington, DC, 247-266.

Neumann, D., E. Zagona, and B. Rajagopalan (2006) A decision support system to manage summer stream temperatures. Journal of the American Water Resources Association 42, 1275-1284.

U.S. Department of Interior (2007) Colorado River Interim Guidelines for Lower Basin Shortages and Coordinated Operations for Lake Powell and Lake Mead, Draft Environmental Impact Statement, Volume 1, Bureau of Reclamation, Boulder City, NV (URL: http://www.usbr.gov/ <http://www.usbr.gov/> lc/ region/programs/strategies/draftEIS/index.html)

Zagona, E., T.J. Fulp, R. Shane, T. Magee, and H. Morgan Goranflo (2001) RiverWare: A Generalized Tool for Complex Reservoir Systems Modeling. Journal of the American Water Resources Association.

Zagona, E., T. Magee, D. Frevert, T. Fulp, M. Goranflo and J. Cotter (2005) RiverWare. In: V. Singh & D. Frevert (Eds.), Watershed Models, Taylor & Francis/CRC Press: Boca Raton, FL, 680pp.

EXECUTIVE SUMMARY

Climate Change Science Program Synthesis and Assessment Product 5.1

challenges and promise of these capabilities and discusses the interaction between users and producers of information (including the role, measurement, and communication of uncertainty and confidence levels associated with decision-support outcomes and their related climate implications).

Earth information—the diagnostics of Earth's climate, water, air, land, and other dynamic processes—is essential for our understanding of humankind's relationship to our natural resources and our environment. Earth information can inform our scientific knowledge, our approach to resource and environmental management and regulation, and our stewardship of the planet for future generations. New data sources, new ancillary and complementary technologies in hardware and software, and ever-increasing modeling and analysis capabilities characterize the current and prospective states of Earth science and are a harbinger of its promise. A host of Earth science data products is enabling a revolution in our ability to understand climate and its anthropogenic and natural variations. Crucial to this relationship, however, is understanding and improving the integration of Earth science information in the activities that support decisions underlying national priorities, ranging from homeland security and public health to air quality and natural resource management.

Also crucial is the role of Earth information in improving our understanding of the processes and effects of climate as it influences or is influenced by actions taken in response to national priorities. Global change observations, data, forecasts, and projections are integral to informing climate science.

The Synthesis and Assessment Product (SAP), "Uses and Limitations of Observations, Data, Forecasts, and Other Projections in Decision Support for Selected Sectors and Regions" (SAP 5.1), examines the current and prospective contributions of Earth science information in decision-support activities and their relationship to climate change science. The SAP contains a characterization and catalog of observational capabilities in a selective set of decision-support activities. It also contains a description of the

Decision-Support Tools and Systems

In 2002, the National Aeronautics and Space Administration (NASA) formulated a conceptual framework in the form of a flow chart (Figure ES-1) to characterize the link between Earth science data and their potential contribution to resource management and public policy. The framework begins with Earth observations, including measurements made in situ and from airborne and space-based instruments. These data are inputs into Earth system models that simulate the dynamic processes of land, the atmosphere, and the oceans. These models lead, in turn, to predictions and forecasts to inform decision-support tools (DST).

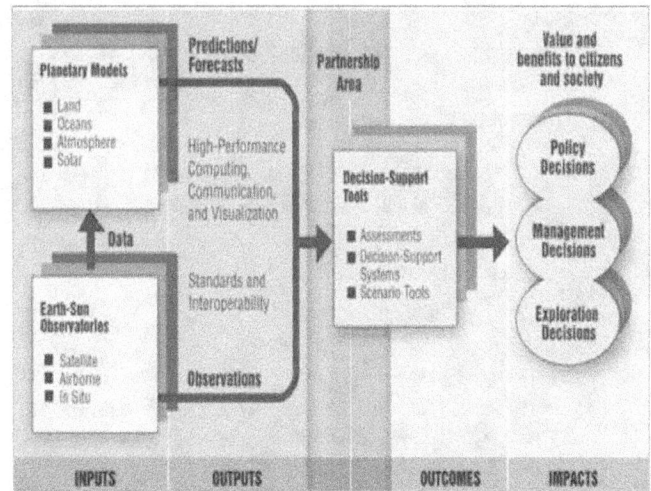

Figure ES-1 The flow of information associated with decision support in the context of variability and change in climate and related systems (Source: Climate Change Science Program [CCSP] Product 5.1 Prospectus, Appendix D)

In this framework, DSTs are typically computer-based models assessing such phenomena as resource supply, the status of real-time events (e.g., forest fires and flooding), or relationships among environmental conditions and other scientific metrics (i.e., water-borne disease vectors and epidemiological data). These tools use data, concepts of relations among data, and analysis functions to allow analysts to build relationships (including spatial, temporal, and process-based) among different types of data, merge layers of data, generate model outcomes, and make predictions or forecasts. Decision-support tools are an element of the broader decision making context or Decision-Support System (DSS). DSSs include not just computer tools but the institutional, managerial, financial, and other constraints involved in the decision-making process.

The outcomes in these decision frameworks are intended to enhance our ability to manage resources (management of public lands and measurements for air quality and other environmental regulatory compliance) and evaluate policy alternatives (as promulgated in legislation or regulatory directives) affecting local, state, regional, national, or even international actions. To be exact, for a variety of reasons, many decisions are not based on data or models. In some cases, formal modeling is not appropriate, timely, or feasible for all decisions. But among decisions that are influenced by this information, the flow chart (Figure ES-1) characterizes a systematic approach for science to be connected to decision processes.

For purposes of providing an organizational framework, the CCSP provides additional description of decision support:

> *In the context of activities within the CCSP framework, decision-support resources, systems, and activities are climate-related products or processes that directly inform or advise stakeholders in order to help them make decisions. These products or processes include analyses and assessments, interdisciplinary research, analytical methods (including scenarios and alternative analysis methodologies), model and data product development, communication, and operational services that provide timely and useful information to decision makers, including policymakers, resource managers, planners, government officials, and other stakeholders.* "Our Changing Planet," CCSP Fiscal Year 2007, Chapter 7, p. 155

Our Approach

Our approach to this SAP has involved two overall tasks. The first task defines and describes an illustrative set of DSTs in areas selected from a number of areas deemed nationally important by NASA and also included in societal benefit areas identified by the intergovernmental Group on Earth Observations (GEO) in leading an international effort to build a Global Earth Observation System of Systems (GOESS) (see Tables ES-1 and ES-2).

Table ES-1 List of NASA National Applications Areas (Appendix B, CCSP SAP 5.1 Prospectus)

Nationally Important Applications	Nationally Important Applications
Agricultural Efficiency	Ecological Forecasting
Air Quality	Energy Management
Aviation	Homeland Security
Carbon Management	Invasive Species
Coastal Management	Public Health
Disaster Management	Water Management

Table ES-2 Societal benefit areas identified by the Group on Earth Observations (GEO) for the Global Earth Observations System of Systems (GEOSS) (http://www.earthobservations.org/about/about_GEO.html) (accessed May 2007)

GEOSS Socio-Benefit Area Keywords	GEOSS Socio-Benefit Area Descriptions
Health	Understanding environmental factors affecting human health and well being
Disasters	Reducing loss of life and property from natural and human-induced disasters
Forecasts	Improving weather information, forecasting, and warning
Energy	Improving management of energy resources
Water	Improving water resource management through better understanding of the water cycle
Climate	Understanding, assessing, predicting, mitigating, and adapting to climate variability and change
Agriculture	Supporting sustainable agriculture and combating desertification
Ecology	Improving the management and protection of terrestrial, coastal, and marine ecosystems

The areas we have chosen as our case studies are air quality, agricultural efficiency, energy management, water management, and public health. As required by the SAP 5.1 Prospectus, in the case studies, we:

- Explain the observational capabilities that are currently or potentially used in these tools;

- Identify the agencies and organizations responsible for their development, operation, and maintenance;

- Characterize the nature of interaction between users and producers of information in delivering accessing and assimilating information;

- Discuss sources of uncertainty associated with observational capabilities and the decision tools and how they are conveyed in decision-support context and to decision makers; and

- Describe relationships between the decision systems and global change information, such as whether the tools at present contribute to (or in the future could contribute to) climate-related predictions or forecasts.

Because our purpose in this first task is to offer case studies by way of illustration rather than a comprehensive treatment of all DSTs in all national applications, in our second task, we have taken steps to catalog other DSTs that use or may use, or that could contribute to, forecasts and projections of climate and global change. The catalog is a first step toward an ever-expanding inventory of existing and emerging DSTs. The catalog will be maintained online for community input, expansion, and updating to provide a focal point for information about the status of DSTs and how to access them.

The information in this report is largely from published literature and interviews with the sponsors and stakeholders of the decision processes, as well as publications by and interviews with the producers of the scientific information used in the tools.

Our Case Studies

We characterize the following DSTs:

1. The Production Estimate and Crop Assessment Division (PECAD) and its Crop Condition Data Retrieval and Evaluation (CADRE) system of the United States (U.S.) Department of Agriculture (USDA), Foreign Agricultural Service (FAS). PECAD/CADRE is the world's most extensive and longest running (over two decades) operational user of remote sensing data for evaluation of worldwide agricultural productivity.

2. The Community Multiscale Air Quality (CMAQ) modeling system of the U.S. Environmental Protection Agency (EPA). CMAQ is a widely used, U.S. continental/regional/urban-scale air quality DST.

3. The Hybrid Optimization Model for Electric Renewables (HOMER), a micropower optimization model of the U.S. Department of Energy's National Renewable Energy Laboratory (NREL). HOMER is used around the world to optimize deployment of renewable energy technologies.

4. Decision-Support System to Prevent Lyme Disease (DDSPL) of the U.S. Centers for Disease Control and Prevention (CDC) and Yale University. DDSPL seeks to prevent the spread of the most common vector-borne disease, Lyme disease, of which there are tens of thousands of cases annually in the U.S.

5. RiverWare, a system developed by the University of Colorado-Boulder's Center for Advanced Decision Support for Water and Environmental Systems (CADSWES) in collaboration with the Bureau of Reclamation, Tennessee Valley Authority, and the Army Corps of Engineers. RiverWare is a hydrologic or river basin modeling system that integrates features of reservoir systems, such as recreation, navigation, flood control, water quality, and water supply, in a basin management tool with power system economics to provide basin managers and electric utilities a method of planning, forecasting, and scheduling reservoir operations.

Taken together, these DSTs demonstrate a rich variety of applications of observations, data, forecasts, and other predictions. In four of our studies, agricultural efficiency, air quality, water management, and energy management, the DSTs have become well established as a basis for public policy decision making. In the case of public health, our lead author points out reasons why direct applications of Earth observations to public health have tended to lag behind these other applications and, thus, is a relatively new application area. He also reminds us that management of air quality, agriculture, water, and energy—in and of themselves—have implications for the quality of public health. The DST he selects is a new, emerging tool intended to assist in prevention of the spread of infectious disease.

Our selection also varies in the geographic breadth of application, illustrating how users of these tools tailor them to relevant regions of analysis and how, in some cases, the geographic coverage of the tools carries over to their requirements for observations. For instance, PECAD/CADRE is used for worldwide study of agricultural productivity and has data requirements of wide geographic scope; HOMER can be used for renewable energy optimization throughout the world; and DDSPL focuses on the eastern, upper Midwest, and West Coast portions of the U.S. CMAQ is used to predict air quality for the contiguous U.S. as well as regions and urban locales. RiverWare provides basin managers and electric utilities with a method of planning, forecasting, and scheduling reservoir operations.

With the exception of DDSPL, none of the DSTs we considered for potential selection, nor those we discuss in this report, have to date made extensive use of climate change information and predictions or have been used to study the effect of a changing climate. However, in all cases, the developers and users of these DSTs fully recognize their applicability to climate change science. In the discussion of the five DSTs presented in this SAP, the authors describe how climate data and/or predictions might be used in these DSTs so that long-range decisions and planning might be accomplished, provided that good quality information and predictions can be ascertained.

Overview of the Chapters

In the Introduction, we provide the rationale for the SAP and a brief overview of the chapters that follow. In the chapters that follow the Introduction, we describe the DST and its data sources, highlight potential uses as well as limits of the DSTs, note sources of uncertainty in using the tools, and finally, discuss the link between the DST and climate change and variability. After our summary, we offer general observations about similarities and differences among the studies.

AGRICULTURAL EFFICIENCY

PECAD of the USDA's FAS uses remote-sensing data for evaluation of worldwide agricultural productivity. PECAD supports the FAS mission to collect and analyze global crop intelligence information and provide periodic estimates used to inform official USDA forecasts for the agricultural market, including farmers; agribusiness; commodity traders and researchers; and federal, state, and local agencies. PECAD is often referred to as PECAD/CADRE with one of its major automated components known as the CADRE geospatial database management system. Of all the DSTs we consider in this report, CADRE has the oldest pedigree as the operational outcome of two early, experimental earth observation projects during the 1970s and 1980s: the Large Area Crop Inventory Experiment (LACIE) and the Agriculture and Resources Inventory Surveys through Aerospace Remote Sensing (AGRISTARS).

Sources of data for CADRE include a large number of weather and other Earth observations from U.S.,

European, Japanese, and commercial systems. PECAD combines these data with crop models, a variety of Geographic Information System (GIS) tools, and a large amount of contextual information, including official government reports, trade and news sources, and on-the-ground reports from a global network of embassy attachés and regional analysts.

Potential future developments in PECAD/CADRE could include space-based observations of atmospheric carbon dioxide (CO_2) measurements and measurement of global sea surface salinity to improve the understanding of the links between the water cycle, climate, and oceans. Other opportunities for enhancing PECAD/CADRE include improvements in predictive modeling capabilities in weather and climate.

One of the largest technology gaps in meeting PECAD requirements is the practice of designing Earth observation systems for research rather than operational use, limiting the capability of PECAD/CADRE to rely on data sources from non-operational systems. PECAD analysts require input data that are collected over long time periods, implying the use of operational systems that ensure continuous data streams and that minimize vulnerability to component failure through redundancy.

Sources of uncertainty can arise at each stage of analysis, from the accuracy of data inputs to the assumptions in modeling. PECAD operators have been able to benchmark, validate, verify, and then selectively incorporate additional data sources and automated decision tools by way of detailed engineering reviews. Another aspect of resolving uncertainty in PECAD is the extensive use of a convergence methodology to assimilate information from regional field analysts and other experts. This convergence of evidence analysis seeks to reconcile various independent data sources to achieve a level of agreement to minimize estimate error.

The relationship between climate and agriculture is complex as agriculture is influenced not only by a changing climate but by agricultural practices themselves, which are contributory to climate change (e.g., in affecting land use and influencing carbon fluxes). At present, PECAD is not directly used to address these dimensions of the climate-agriculture interaction. However, many of the data inputs for PECAD are climate-related, thereby enabling PECAD to inform the understanding of agriculture as a "recipient" of climate-induced changes. For instance, observing spatial and geographic trends in the output measures from PECAD can contribute to understanding how the agricultural sector is responding to a changing climate. Likewise, trends in PECAD's measures of the composition and production of crops could shed light on the agricultural sector as a "contributor" to climate change (for instance, in terms of greenhouse gas emissions or changes in soil that may affect the potential for agricultural soil carbon sequestration). The results produced by PECAD may also be influenced by climate-induced changes in land use. In addition, the influences may work in the other direction. The changes in the results overtime may be a barometer of land use changes, such as the conversion from food production to biomass fuel production.

AIR QUALITY

The EPA CMAQ modeling system has been designed to approach air quality by including state-of-the-science capabilities for modeling tropospheric ozone, fine particles, toxics, acid deposition, and visibility degradation. CMAQ is used to guide the development of air quality regulations and standards and to create state implementation plans for managing air emissions. CMAQ also can be used to evaluate longer-term as well as short-term transport from localized sources and to perform simulations using downscaled regional climate from global climate change scenarios.

The CMAQ modeling system contains three types of modeling components: a meteorological modeling system for the description of atmospheric states and motions, emission models for man-made and natural emissions that are injected into the atmosphere, and a chemistry-transport modeling system for simulation of the chemical transformation and fate. Inputs for CMAQ and their associated regional meteorological model, mesoscale model version 5 (MM5) can include, but are not limited to, the comprehensive output from a general circulation model, anthropogenic and biogenic emissions, description of wildland fires, land use and demographic changes, and meteorological and atmospheric chemical species measurements by in-situ and remote-sensing platforms, including satellites and aircraft.

A major source of uncertainty for CMAQ has been the establishment of initial conditions. The default initial conditions and lateral boundary conditions in CMAQ are provided under the assumption that after spin-up of the model, they no longer play a role, and in time, surface emissions govern the air quality found in the lower troposphere. However, it has been shown that the effects of the lateral boundary conditions differ for different latitudes, altitudes, and seasons. Other sources of uncertainty in CMAQ are due to uncertainties in the emissions inventory, limitations in science parameterizations, and modeling difficulties produced by such factors as spatial resolution.

CMAQ can be used to answer many climate-change, air quality-related questions. In order to accomplish this, CMAQ will require information on such factors as greenhouse gases, global warming, population growth, land use changes, new emission controls being implemented, and the availability of new energy sources to replace the existing high-carbon sources. Scenarios can be chosen either to study potential impacts or to estimate the range of uncertainties of the predictions. Global air quality models must be combined with CMAQ to resolve the effects of climate change on air quality. CMAQ would be used to downscale the coarse-scale predictions of the global model to regional or local scale.

ENERGY MANAGEMENT

HOMER is a micropower optimization model of the U.S. Department of Energy's NREL. HOMER is capable of calculating emission reductions enabled by replacing diesel-generating systems with renewable energy systems in a micro-grid or grid-connected configuration. HOMER helps the user design grid-connected and off-grid renewable energy systems by performing a wide range of design scenarios. HOMER can be used to address questions such as:

- Which technologies are most cost effective?

- What happens to the economics if the project's costs or loads change?

- Is the renewable energy resource adequate for the different technologies being considered to meet the load?

HOMER does this by finding the least-cost combination of components that meet electrical and thermal loads.

The Earth observation information serving as input to HOMER is centered on wind and solar resource assessments derived from a variety of sources. Wind data include surface and upper air station data, satellite-derived ocean and ship wind data, and digital terrain and land cover data. Solar resource data include surface cloud, radiation, aerosol optical depth (AOD), and digital terrain and land cover data from both in-situ and remote-sensing sources.

All of the input data for HOMER can have a level of uncertainty attached to them. HOMER allows the user to perform sensitivity tests on one or more variables and has graphical capabilities to display these results to inform decision makers. As a general rule, the error in estimating the performance of a renewable energy system over a year is roughly linear to the error in the input resource data.

One of the largest challenges in HOMER is the absence of direct or in-situ solar and wind resource measurements at specific locations to which HOMER is applied. In addition, in many cases, values are not based on direct measurement at all but are approximations based on the use of algorithms to convert a signal into the parameter of interest as is the case with most satellite-derived data products. For example, satellite-derived ocean wind data are not based on direct observation of the wind speed above the ocean surface but are derived from an algorithm that infers wind speed based on wave height observations. Observations of AODs (for which considerable research is underway) can be complicated by irregular land-surface features that place limitations on the application of algorithms for satellite-derived measures.

For renewable energy resource mapping, improved observations of key weather parameters (for instance, wind speed and direction at various heights above the ground, particularly at the hub height of wind energy turbine systems; over the open oceans at higher and higher spatial resolutions; and improved ways of differentiating snow cover and bright reflecting surfaces from clouds) will be of value to the renewable energy community. New, more accurate methods of related parameters, such as AOD, would also improve the resource data.

The relationship between HOMER and global change information is largely by way of the dependence of renewable energy resource input measurements on weather and local climate conditions. Although HOMER was not designed to be a climate-related management decision-making tool, by optimizing the mix of hybrid renewable energy technologies for meeting load conditions, HOMER can enable users to respond to climate change and variability in their energy management decisions. HOMER could be used to evaluate how renewable energy systems can be used cost effectively to displace fossil fuel-based systems.

PUBLIC HEALTH

The DDSPL is operated by the U.S. CDC and Yale University to address questions related to the likely distribution of Lyme disease east of the 100th meridian, where most cases occur. Lyme disease is the most common vector-borne disease in the U.S., with tens of thousands of cases annually. Most human cases occur in the Eastern and upper Midwest portions of the U.S., although there is a secondary focus along the West Coast. Vector-borne diseases are those in which parasites (virus, bacterium, or other micro-organism) are transmitted among people or from wildlife to people by insects or arthropods (as vectors, they do not themselves

cause disease). The black-legged tick is typically the carrier of the bacteria causing Lyme disease.

Early demonstrations during the 1980s showed the utility of Earth observations for identifying locations and times that vector-borne diseases were likely to occur, but growth of applications has been comparatively slow. Earth-observing instruments have not been designed to monitor disease risk; rather, data gathered from these platforms are "scavenged" for public health risk assessment. DDSPL uses satellite data and derived products, such as land cover together with meteorological data and census data, to characterize statistical predictors of the presence of black-legged ticks. The model is validated by field surveys. The DDSPL is thus a means of setting priorities for the likely geographic extent of the vector; the tool does not at present characterize the risk of disease in the human population.

Future use of DDSPL partly depends on whether the goal of disease prevention or the goal of treatment drives public health policy decisions. In addition, studies have shown that communication to the public about the risk in regions with Lyme disease often fails to reduce the likelihood of infection. The role of improved Earth science data is unclear in terms of improving the performance of DDSPL because, at present, the system has a level of accuracy deemed "highly satisfactory." Future use may instead require a model of sociological/behavioral influences among the population.

Standard statistical models and in-field validation are used to assess the uncertainty in decision making with DDSPL. The accuracy of clinical diagnoses also influences the ultimate usefulness of DDSPL as an indicator tool to characterize the geographic extent of the vectors.

The DDSPL is one of the few public health DSTs that has explicitly evaluated the effects of climate variability. Using outputs of a Canadian climate change model, study has shown that with warming global mean temperatures predicted by the year 2050 to 2080, the geographic range of the tick vector will decrease at first, with reduced presence in the southern boundary, and then expand into Canada and the central region of North America where it now absent. The range also moves away from population concentrations.

WATER MANAGEMENT

RiverWare was developed and is maintained by CADSWES in collaboration with the Bureau of Reclamation, Tennessee Valley Authority, and the Army Corps of Engineers. It is a river basin modeling system that integrates features of reservoir systems, such as recreation, navigation, flood control, water quality, and water supply in a basin management tool, with power system economics to provide basin managers and electric utilities with a method of planning, forecasting, and scheduling reservoir operations. RiverWare uses an object-oriented software engineering approach in model development. The object-oriented software-modeling strategy allows computational methods for new processes, additional controllers for providing new solution algorithms, and additional objects for modeling new features to be added easily to the modeling system. RiverWare is data intensive in that a specific river/reservoir system and its operating policies must be characterized by the data supplied to the model. This allows the models to be modified as new features are added to the river/reservoir system and/or new operating policies are introduced. The data-intensive feature allows the model to be used for water management in most river basins.

RiverWare is menu driven through a graphical user interface (GUI). The basin topology is developed through the selection of a reservoir, reach, confluence, and other necessary objects and by entering the data associated with each object manually or through importing files. Utilities within RiverWare provide a means to automatically execute many simulations, to access data from external sources, and to export model results. Users also define operating policies through the GUI as system constraints or rules for achieving system management goals (e.g., related to flood control, water supplies, water quality, navigation, recreation, and power generation). The direct use of Earth observations in RiverWare is limited. Unlike traditional hydrologic models that track the transformation of precipitation (e.g., rain and snow) into soil moisture and streamflow, RiverWare uses supplies of water to the system as input data. These data are derived from a hydrologic model where direct use of earth observations can be and have been made. Application of RiverWare is limited by the specific implementation defined by the user and by the quality of the input data. It has tremendous flexibility in the kinds of data it can use, but long records of data are required to overcome the issue of data non-stationarity.

The reliability of observations for driving hydrologic models that may provide input to RiverWare is a major source of uncertainty for RiverWare, as is the hydrologic models themselves. The major sources of uncertainty in RiverWare include (1) errors in estimates of precipitation, soil moisture, evapotranspiration (ET), and human manipulation of stream flows; (2) difficulties in reliable and timely processing of data into usable forms; (3) mismatches in space and time scales between atmospheric and land surface processes and their models;

(4) incomplete description of physical processes in land surface and hydrologic models, and (5) uncertain error characteristics of the outputs of atmospheric, land surface, and hydrologic models.

Decision makers recognize that with a changing climate, mid- and long-range planning for the sustainability of water resources is an absolute requirement. RiverWare is capable of supporting climate-related water resources management decisions. The specific application of RiverWare in the context of mid- or long-range planning for a specific river basin will reflect whether decisions may rely on global change information. For mid-range planning of reservoir operations, characterization and projections of interannual and decadal-scale climate variability (e.g., monitoring, understanding, and predicting interannual climate phenomena such as the El Nino-Southern Oscillation) are important. For long-term planning, global warming has moved from the realm of speculation to general acceptance. The impacts of global warming on water resources and their implications for management have been a major focus in the assessments of climate change. The estimates of potential impacts of climate change on precipitation have been inconclusive, leading to increasing uncertainty about the reliability of future water supplies. Uncertainty in climate predictions and in watershed behavior, river hydraulics, and management policies in the future, as well as poor monitoring of human impacts on natural stream flow will produce significant uncertainties in long-term planning and design applications using RiverWare.

General Observations

Application of all of the DSTs involves a variety of input data types, all of which have some degree of uncertainty in terms of their accuracy. The amount of uncertainty associated with resource data can depend heavily on how the data are obtained. Quality in-situ measurements of wind and solar data suitable for application in HOMER can have uncertainties of less than ± 3% of true value; however, when estimation methods are required, such as the use of Earth observations, modeling, and empirical techniques, uncertainties can be as much as ± 10% or more. The DSTs address uncertainty by allowing users to perform sensitivity tests on variables. With the exception of HOMER, a significant amount of additional traditional on-the-ground reports are a critical component. In the case of PECAD/CADRE, uncertainty is resolved in part by extensive use of a convergence methodology to assimilate information from regional field analysts and other experts. This brings a large amount of additional information to PECAD/CADRE forecasts, well beyond the automated outputs of DSTs. In RiverWare, streamflow and other hydrologic variables respond to atmospheric factors, such as precipitation, and obtaining quality precipitation estimates is a formidable challenge, especially in the western U.S. where orographic effects produce large spatial variability and where there is a scarcity of real-time precipitation observations and poor radar coverage.

In terms of their current or prospective use of climate change predictions or forecasts as DST inputs, or the contributions of DST outputs to understanding, monitoring, and responding to a changing climate, the status is mixed. DDSPL is one of the few public health decision-support tools that has explicitly evaluated the potential impact of climate change scenarios on an infectious disease system. None of the other DSTs at present is directly integrated with climate change measurements, but all of them can and may in the future take this step. PECAD/CADRE's assessment of global agricultural production will certainly be influenced by reliable observations and forecasts of climate change and variability as model inputs, just as the response of the agricultural sector to a changing climate will feed back into PECAD/CADRE production estimates. HOMER's renewable energy optimization calculations will be directly affected by climate-related changes in renewable energy resource supplies and will enhance our ability to adapt to climate-induced changes in energy management and forecasting. Air quality will likely be affected by global climate change. The capability of CMAQ to predict those affects is conditional on acquiring accurate predictions of the meteorology under the climate change conditions that will take place in the U.S. and accurate emission scenarios for the future. Given these inputs to CMAQ, reliable predictions of the air quality and their subsequent health affects can be ascertained. It was noted that there is great difficulty in integrating climate change information into RiverWare and other such water management models. The multiplicity of scenarios and vague attribution of their probability for occurrence, which depends on feedback among social, economic, political, technological, and physical processes, complicates conceptual integration of climate change impacts assessment results in a practical water management context. Furthermore, the century time scales of climate change exceed typical planning and infrastructure design horizons in water management.

Audience and Intended Use

The CCSP SAP 5.1 Prospectus describes the audience and intended use of this report:

This synthesis and assessment report is designed to serve decision makers and stakeholder communities interested in using global change information resources in policy, planning, and other practical uses. The goal is to provide useful information on climate change research products that have the capacity to inform decision processes. The report will also be valuable to the climate change science community because it will indicate types of information generated through the processes of observation and research that are particularly valuable for decision support. In addition, the report will be useful for shaping the future development and evaluation of decision-support activities, particularly with regard to improving the interactions with users and potential users.

There are a number of national and international programs focusing on the use of Earth observations and related prediction capacity to inform decision-support tools (see Table ES-3, "Related National and International Activities"). These programs both inform and are informed by the CCSP and are recognized in the development of this product. (CCSP SAP 5.1, Prospectus for "Uses and Limitations of Observations, Data, Forecasts, and Other Projections in Decision Support for Selected Sectors and Regions," 28 February 2006)

Table ES-3 References to Related National and International Activities (Source: Appendix C, CCSP SAP 5.1 Prospectus)

Priority	National	International
Climate Change	CCSP and Climate Change Technology Program	Intergovernmental Panel on Climate Change and World Climate Research Programme
Global Earth Observations	National Science and Technology Council Committee on Environment and Natural Resources Subcommittee U.S. GEO	GEO
Weather	U.S. Weather Research Program	World Meteorological Organization
Natural Hazards	National Science and Technology Council (NSTC) Committee on Environment and Natural Resources Research (CENR) Subcommittee on Disaster Reduction	International Strategy for Disaster Reduction
Sustainability	NSTC CENR Subcommittee on Ecosystems	World Summit on Sustainable Development
E-Government	Geospatial One-Stop and the Federal Geographic Data	World Summit on the Information Society

INTRODUCTION

This Synthesis and Assessment Product (SAP), "Uses and Limitations of Observations, Data, Forecasts, and Other Projections in Decision Support for Selected Sectors and Regions" (SAP 5.1), examines the current and prospective contribution of Earth science information/data in decision-support activities and their relationship to climate change science. The SAP contains a characterization and catalog of observational capabilities in an illustrative set of decision-support activities. It also contains a description of the challenges and promises of these capabilities and discusses the interaction between users and producers of information, including the role, measurement, and communication of uncertainty and confidence levels associated with decision-support outcomes and their related climate implications.

The organizing basis for the chapters in this SAP is the decision-support tools (DST), which are typically computer-based models assessing such phenomena as resource supply, the status of real-time events (e.g., forest fires and flooding), or relationships among environmental conditions and other scientific metrics (for instance, water-borne disease vectors and epidemiological data). These tools use data, concepts of relations among data, and analysis functions to allow analysts to build relationships (including spatial, temporal, and process-based) among different types of data, merge layers of data, generate model outcomes, and make predictions or forecasts. DSTs are an element of the broader decision-making context—the Decision-Support System (DSS). DSSs include not just computer tools but also the institutional, managerial, financial, and other constraints involved in decision making.

Our approach to this SAP is to define and describe an illustrative set of DSTs in areas selected from topics deemed nationally important and included in societal benefit areas identified by the intergovernmental Group on Earth Observations (GEO) in leading an international effort to build a Global Earth Observation System of Systems (GEOSS). The areas we have chosen as our focus are air quality, agricultural efficiency, energy management, water management, and public health. The DSTs we characterize are:

1. The Production Estimate and Crop Assessment Division (PECAD) and its Crop Condition Data Retrieval and Evaluation (CADRE) system of the United States (U.S.) Department of Agriculture (USDA), Foreign Agricultural Service (FAS). PECAD/CADRE is the world's most extensive and longest running (over two decades) operational user of remote sensing data for evaluation of worldwide agricultural productivity.

2. The Community Multiscale Air Quality (CMAQ) modeling system of the U.S. Environmental Protection Agency (EPA). CMAQ is a widely used, U.S. continental/regional/urban-scale air quality DST.

3. The Hybrid Optimization Model for Electric Renewables (HOMER), a micropower optimization model of the U.S. Department of Energy's National Renewable Energy Laboratory (NREL). HOMER is used around the world to optimize deployment of renewable energy technologies.

4. The DSS to Prevent Lyme Disease (DSSPL) of the U.S. Centers for Disease Control and Prevention (CDC) and Yale University. DSSPL seeks to prevent the spread of the most common vector-borne disease, Lyme disease, of which there are tens of thousands of reported cases annually in the U.S.

5. RiverWare, a system developed by the University of Colorado-Boulder's Center for Advanced Decision Support for Water and Environmental Systems (CADSWES) in collaboration with the Bureau of Reclamation, Tennessee Valley Authority, and the Army Corps of Engineers, is a hydrologic or river basin modeling system that integrates features of reservoir systems, such as recreation, navigation, flood control, water quality, and water supply, in a basin management tool with power system economics to provide basin managers and electric utilities with a method of planning, forecasting, and scheduling reservoir operations.

Taken together, these DSTs demonstrate a rich variety of applications of observations, data, forecasts, and other predictions. In four of our studies—agricultural efficiency, air quality, water management, and energy management—the DSTs have become well established as a basis for public policy decision making. In the case of public health, our lead author points out reasons why direct applications of Earth observations to public health have tended to lag behind these other applications and, thus, is a relatively new applications area. He also reminds us that management of air quality, agriculture, water, and energy—in and of themselves—have implications for the quality of public health. The DST selected for public health is a new and emerging tool intended to assist in the prevention of the spread of infectious disease.

With the exception of DSSPL, none of the DSTs we considered for potential selection, nor those we discuss in this report, have to date made extensive use of climate change information or been used to study the effect of a changing climate. However, in all cases, the developers and users of these DSTs fully recognize their applicability to climate change science. In the discussion of the five DSTs presented in this SAP, the authors describe how reliable climate data and/or predictions might be used in these DSTs so that long-range decisions and planning might be accomplished.

1

CHAPTER

Decision Support for Agricultural Efficiency

Lead Author: Molly K. Macauley

1. Introduction

The efficiency of agriculture has been one of the most daunting challenges confronting mankind in its need to manage natural resources within the constraints of weather, climate, and other environmental conditions. Defined as maximizing output per unit of input, agricultural efficiency reflects a complex relationship among factors of production (including seed, soil, human, and physical capital) and the exogenous influence of nature (such as temperature, sunlight, weather, and climate). The interaction of agricultural activity with the environment creates another source of interdependence (e.g., the effect on soil and water from applications of pesticides, fungicides, and fertilizer). Agricultural production has long been a large component of international trade and of strategic interest as an indicator of the health and security of nations.

The relationship between climate change and agriculture is complex. A changing climate can influence agricultural practices (e.g., climate-induced changes in patterns of rainfall could lead to changes in these practices). Agriculture is not only influenced by a changing climate, but agricultural practices themselves are a contributory factor through emissions of greenhouse gases and influences on fluxes of carbon through photosynthesis and respiration. In short, agriculture is both a contributor to and a recipient of the effects of a changing climate (Rosenzweig, 2003; National Assessment Synthesis Team, 2004).

The use of Earth observations by the agricultural sector has a long history. The Large Area Crop Inventory Experiment (LACIE), jointly sponsored by the United States (U.S.) National Aeronautics and Space Administration (NASA), the U.S. Department of Agriculture (USDA), and the National Oceanic and Atmospheric Administration (NOAA) from 1974 to 1978, demonstrated the potential for satellite observations to make accurate, extensive, and repeated surveys for global crop forecasts. LACIE used observations from the land remote-sensing satellite (Landsat) series of multispectral scanners on sun-synchronous satellites. The Agriculture and Resources Inventory Surveys through Aerospace Remote Sensing (AgRISTARS) followed LACIE and extended the use of satellite observations to include early warning of production changes, inventory and assessment of renewable resources, and other activities (Congressional Research Service, 1983; National Research Council [NRC], 2007; Kaupp et al., 2005). Today, these data are used by agencies of the federal government, commodity trading companies, farmers, relief agencies, other governments, and essentially anyone with an interest in crop production at a global scale.

An approach, among others, to increasing agricultural efficiency is to expand and enhance uses of Earth observation data for (1) policy and resource management decision support, (2) monitoring and measuring climate change affects, and (3) providing policy and resource climate change decision support. The foremost example of the application of Earth observations in agriculture is found in the USDA's crop-monitoring Decision-Support System (DSS), the Production Estimates and Crop Assessment Division (PECAD) of the USDA's Foreign Agricultural Service (FAS). (Reorganization at USDA finds the PECAD functionality, but not the name, residing within the USDA's FAS as part of the Office of Global Analysis, Impact Analysis Division, International Production Assessment). PECAD is now the world's most extensive and longest running (over two decades) operational user of remote-sensing data for evaluation of worldwide agricultural productivity (NASA, 2001). A description of the PECAD DSS, its functionality, its analysis style, how it deals with making decisions under uncertainty, and its future uses form the basis of this chapter.

2. Description of PECAD

The USDA/FAS uses PECAD to analyze global agricultural production and crop conditions affecting planting, harvesting, marketing, commodity export and pricing, drought monitoring, and food assistance. Access to and uses of PECAD are largely by the federal government, rather than state and local governments, as a means of assessing regions of interest in global agricultural production.

PECAD uses satellite data, worldwide weather data, and agricultural models in conjunction with FAS overseas post reports, foreign government official reports, and agency travel observations to support decision making. FAS also works closely with the USDA Farm Service Agency and the Risk Management Agency to provide early warning and critical analysis of major crop events in the U.S. (FAS online crop assessment at http://www. fas.usda.gov/pecad2/crop_assmnt.html, accessed April 2007). FAS seeks to promote the security and stability of the U.S. food supply, improve foreign market access for U.S. agricultural products, provide reports on world food security, and advise the U.S. government on international food aid requirements. FAS bears the

primary responsibility for USDA's overseas activities: market development, international trade agreements and negotiations, and the collection and analysis of statistics and market information. FAS also administers the USDA's export credit guarantee and food aid programs.

PECAD's Crop Condition Data Retrieval and Evaluation (CADRE) database management system, the operational outcome of the LACIE and AgRISTARs projects, was one of the first geographic information systems (GIS) designed specifically for global agricultural monitoring (Reynolds, 2001). CADRE is used to maintain a large satellite imagery archive to permit comparative interpretation of incoming imagery with that of past weeks or years. The database contains multi-source weather data and other environmental data that are incorporated as inputs for models to estimate parameters such as soil moisture, crop stage, and yield. These models also indicate the presence and severity of plant stress or injury. The information from these technologies is used by PECAD to produce, in conjunction with the World Agricultural Outlook Board, official USDA foreign crop production estimates (FAS online crop assessment at http://www.fas.usda.gov/pecad2/crop_assmnt.html, accessed April 2007).

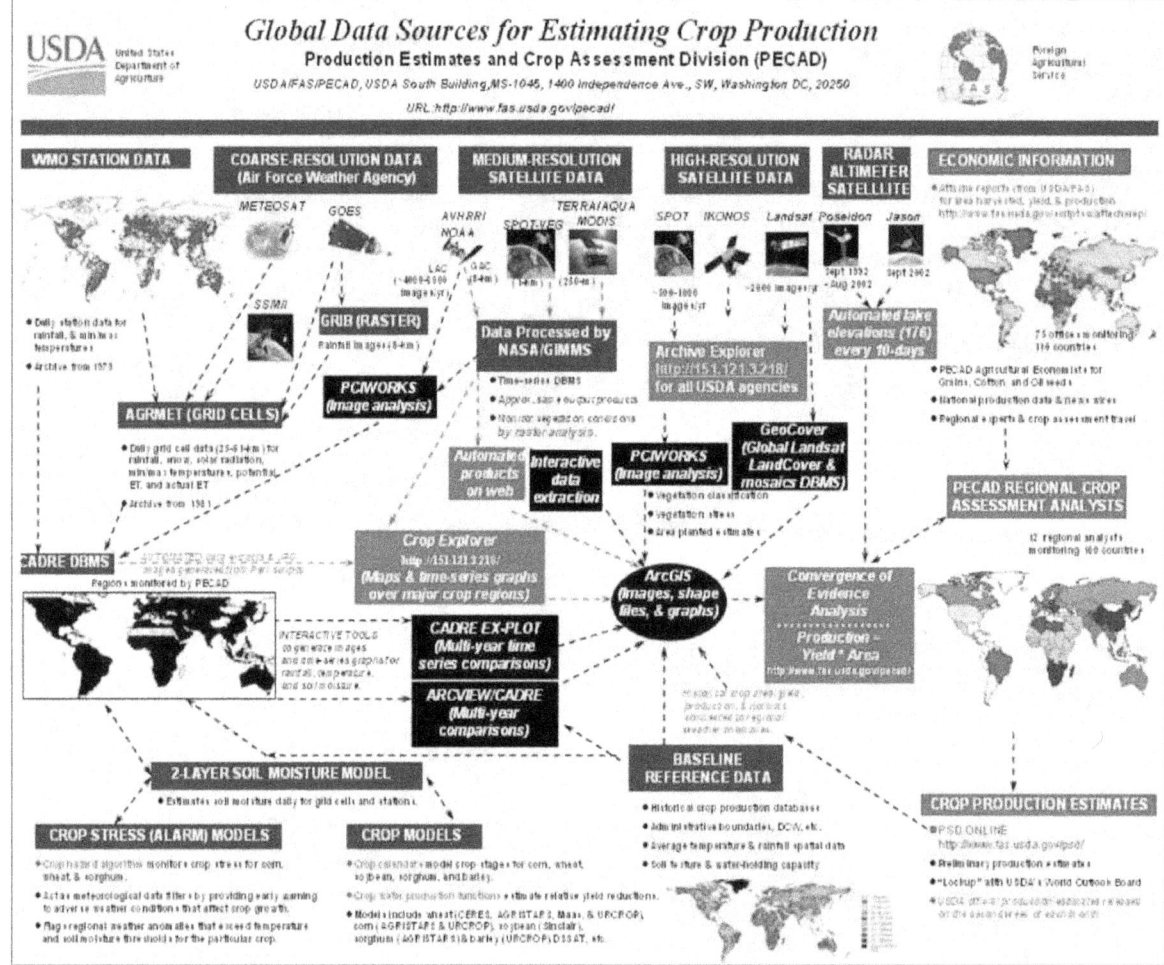

Figure 1-1 The PECAD DSS: Data Sources and DSTs (Source: Kaupp and coauthors, 2005, p. 5)

Figure 1-1 (Kaupp et al., 2005, p. 5) illustrates the global data sources and decision-support tools (DST) for PECAD. The left-hand portion of the figure shows sources of data for the CADRE geospatial DBMS. These inputs include station data from the World Meteorological Organization and coarse resolution data from Meteosat, Scanning Multichannel Microwave Radiometer (SSMR), and Geostationary Satellite (GOES). Meteosat, operated by the European Organization for the Exploitation of Meteorological Satellites, provides visible and infrared weather-oriented imaging. The SSMR and its successor, the Special Sensor Microwave/Imager (SSM/I), are microwave radiometric instruments in the U.S. Air Force Defense Meteorological Satellite program. Additional weather data come from the U.S. GOES program.

Medium resolution satellite data include Advanced Very High Resolution Radiometer (AVHRR)/NOAA, Systeme Pour L'Observation de la Terre (SPOT)-Vegetation (SPOT-VEG), and Terra/Aqua Moderate Resolution Imaging Spectroradiometers (MODIS). AVHRR/NOAA, operated by NOAA, provides cloud cover and land, water, and sea surface temperatures at approximately 1-kilometer (km) spatial resolution. The SPOT supplies commercial optical Earth imagery at resolutions from 2.5 to 20 meters (m); SPOT-Vegetation is a sensor providing daily coverage at 1 km resolution. The NASA MODIS on the Terra and Aqua satellites, part of the U.S. Earth Observation System, show rapid biological and meteorological changes at 250 to 1,000 m spatial resolution every two days. NASA's Global Inventory Modeling and Mapping Studies (GIMMS) group processes data acquired from SPOT and Terra/Aqua MODIS. NASA/GIMMS provides PECAD with a cross-calibrated global time series of Normalized Difference Vegetation Index maps from AVHRR and SPOT-VEG. Moderate-resolution Earth observation data are also used from the U.S. Landsat program.

Sources of high resolution and radar altimeter satellite data include SPOT, IKONOS, Poseidon, and Jason. IKONOS is a commercial Earth imaging satellite providing spatial resolution of 1 and 4 m. Data from Poseidon and its successor, Jason, provide lake and reservoir surface elevation estimates. Poseidon, part of the Ocean Surface Topography Experiment (TOPEX)/ Poseidon mission, and Jason-1, a follow-on mission, are joint ventures between NASA and the Centre National d'Etudes Spatiales using radar altimeters to map ocean surface topography (including sea surface height, wave height, and wind speed above the ocean). These data enable analysts to assess drought or high water-level conditions within some of the world's largest lakes and reservoirs to predict effects on downstream irrigation potential and inform production capacity

estimates (Birkett and Doorn, 2004; Kanarek, 2005). The assimilation of these data into PECAD is described in detail in a recent systems engineering report (NASA, 2004b).

PECAD combines the satellite and climate data, crop models (along the bottom portion of the figure), a variety of GIS tools, and a large amount of contextual information, including official government reports, trade and new sources, and on-the-ground reports from a global network of embassy attachés and regional analysts. The integration and analysis is attained by "convergence of evidence analysis" (Kaupp et al., 2005). This convergence methodology seeks to reconcile various independent data sources to achieve a level of agreement to minimize estimate error (NASA, 2004a).

The crop assessment products indicated along the right-hand side of the PECAD architecture in Figure 1-1 represent the periodic global estimates used to inform official USDA forecasts. These products are provided to the agricultural market, including farmers; agribusiness; commodity traders and researchers; and federal, state, and local agencies. In addition to CADRE, other automated components include two features providing additional types of information. The FAS Crop Explorer (middle of diagram) has been a feature on the FAS Web site since 2002 (Kanarek, 2005). Crop Explorer offers near real-time global crop condition information based on satellite imagery and weather data from the CADRE database and NASA/GIMMS. Thematic maps of major crop growing regions show vegetation health, precipitation, temperature, and soil moisture. Time-series charts show growing season data for agro-meteorological zones. For major agriculture regions, Crop Explorer provides crop calendars and crop areas. Through Archive Explorer, PECAD provides access to an archive of moderate- to high-resolution data, allowing USDA users (access is controlled by user name and password) to search an image database.

3. Potential Future Use and Limits

The most recent enhancements to PECAD/CADRE have included the integration and evaluation of MODIS, TOPEX/Poseidon, and Jason-1 products (NASA, 2006a). Figure 1-2 summarizes the Earth system models, Earth observations data, and the CADRE DBMS and characterizes their outputs. Several planned Earth observations missions anticipated when this image was prepared (indicated in italics) show how PECAD/CADRE could incorporate new opportunities, including those with additional land, atmosphere, and ocean observations. These would include space-based observations of atmospheric carbon dioxide (CO_2) from the Orbiting

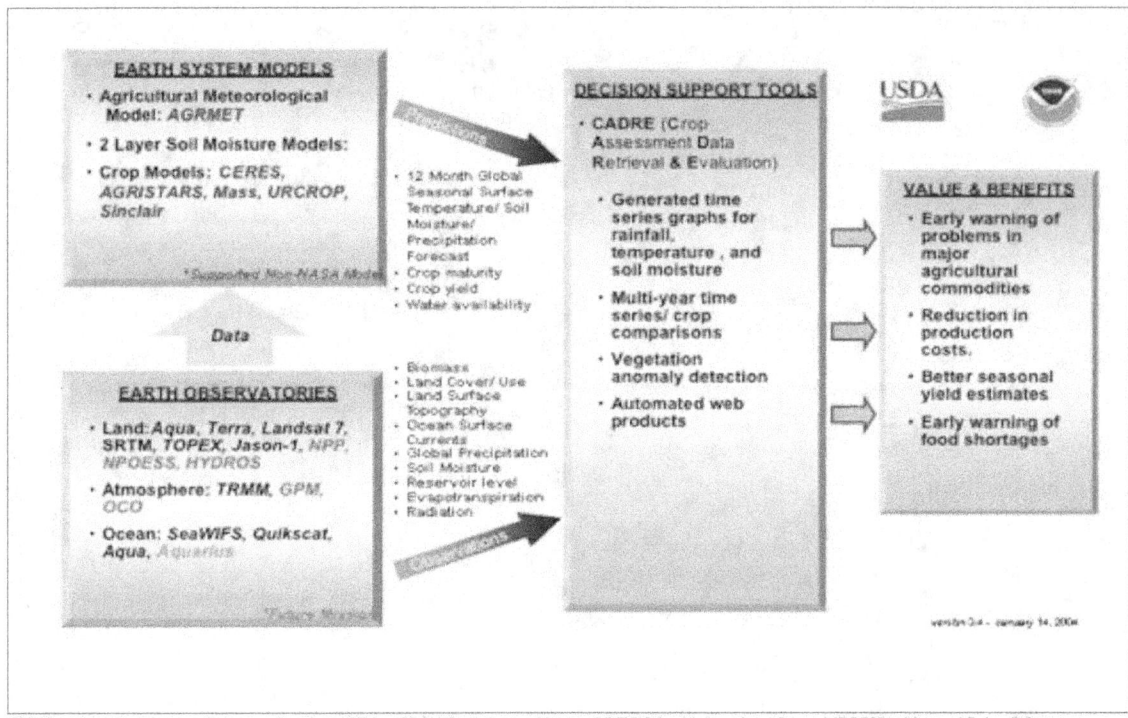

CERES = Crop Environment Resource Synthesis; GPM = Global Precipitation Mission; HYDROS = Hydrosphere State; NPOESS = National Polar-Orbiting Operational Environmental Satellite; NPP = NPOESS Preparatory Project; Quickscat = quick scatterometer; SeaWiFS = Sea-viewing Wide Field-of-view Sensor; SRTM = Shuttle Radar Topology Mission; TRMM = Tropical Rainfall Mapping Mission

Figure 1-2 The PECAD DSS: Earth System Models, Earth Observations, DSTs, and Outputs (Source: NASA, 2006a, p. 32).

Carbon Observatory (OCO) and measurement of global sea surface salinity (Aquarius) to improve understanding of the links between the water cycle, climate, and the ocean. Other opportunities for enhancing PECAD/CADRE could include improvements in predictive modeling capabilities in weather and climate (NASA, 2006a).

In a recent evaluation report for PECAD, NASA has acknowledged that one of the largest technology gaps in meeting PECAD requirements is the design of NASA systems for limited duration research purposes rather than for long-term operational uses (NASA, 2004a). PECAD analysts require long-term continuity for inputs, implying the use of operational systems that ensure continuous data streams over time and that minimize vulnerability to component failure through redundancy. The report also emphasizes that PECAD requires systems that deliver real-time or near real-time data. Many NASA missions have traded timeliness for experimental research or improvements in other properties of the information delivered. Additionally, the report identifies several potential Earth science data streams that have not yet been addressed, including water balance, the radiation budget (including solar and long-wave radiation flux), and elevation, and expresses concern about the potential continuity gap between Landsat 7 and the Landsat Data Continuity Mission.

A 2006 workshop convened at the United Nations Food and Agriculture Organization by the Integrated Global Observations of Land team identified priorities for agricultural monitoring during the next 5 to 10 years as part of the emerging Global Earth Observations System of Systems (GEOSS). In summary, the meeting called for several initiatives including the following (United Nations Food and Agriculture Organization, 2006):

1. The need for an international initiative to fill the data gap created by the malfunction of Landsat 7;

2. A system to collect cloud-free, high resolution (10 to 20 m) visible, near-infrared, and shortwave infrared observations at 5- to 10-day intervals;

3. Workshops on global agricultural data coordination and on integrating satellite and in-situ observations;

4. An inventory and evaluation of existing agro-meteorological datasets to identify gaps in terrestrial networks, the availability of data, and validation and quality control in order to offer specific recommendations to the World Meteorological Organization to improve its database;

5. Funding to support digitizing, archiving, and dissemination of baseline data; and

6. An international workshop within the GEOSS framework to develop a strategy for "community of practice" for improved global agricultural monitoring.

A recent study by the NRC of the use of land remote sensing expressed additional concerns about present limits on the usefulness of Earth observations in agricultural assessment (NRC, 2007). These include data integration, communication of results, and the capacity to use and interpret data. Specifically, the NRC identified these concerns:

1. Inadequate integration of spatial data with socioeconomic data (locations and vulnerabilities of human populations and access to infrastructure) to provide information that is effective in generating response strategies to disasters or other factors influencing access to food or impairing agricultural productivity;

2. A lack of communication between remote-sensing mission planners, scientists, and decision makers to ascertain what types of information enable the most effective food resource management; and

3. Shortcomings in the acquisition, archiving, and access to long-term environmental data and development of capacity to interpret these data, including maintaining continuity of satellite coverage over extended timeframes, providing access to affordable data, and improving the capacity to interpret data.

4. Uncertainty

Two aspects of PECAD provide a means of validation and verification of crop assessments. One is the maturity of PECAD as a DSS. Over the years, PECAD has been able to benchmark, validate, verify, and then selectively incorporate additional data sources and automated decision tools. An example of the systems engineering review associated with a decision to incorporate Poseidon and Jason data, for example, is offered in a detailed NASA study (NASA, 2004b).

Another example demonstrates how data product accuracy, delivery, and coverage are tested through validation and verification during the process of assimilating new data sources and how they ascertain the extent to which different data sources corroborate model outputs (Kaupp et al., 2005). Essential considerations included enhanced repeatability of results, increased accuracy, and increased throughput speed.

Another significant aspect of resolving uncertainty in PECAD is its extensive use of a convergence methodology to assimilate information from regional field analysts and other experts. PECAD seeks to provide accurate and timely estimates of production yet must accommodate physical and biological influences (e.g., weather or pests), the fluctuations in agricultural

markets, and developments in public policy impacting the agricultural sector (Kaupp et al., 2005). The methodology brings a large amount of additional information to the PECAD forecasts, well beyond the automated outputs of the DSTs. This extensive additional analysis may not fully correct for, but certainly mitigates, the uncertainty inherent in the data and modeling at the early stages. Figure 1-3, a simplified version of Figure 1-1, shows the step represented by the analyses that take place during this convergence of information in relation to the outputs obtained from the DSTs and their data inputs. Figure 1-4 further describes the nature of information included in the convergence methodology in addition to the outputs of the data and automated DSTs. Official reports, news reports, field travel, and attaché reports are additional inputs at this stage. The process is described as one in which, "while individual analysts reach their conclusions in different ways, giving different weight to various inputs, analysts join experts from the USDA's Economic Research Service and National Agricultural Statistics Service once a month in a 'lock-up.' In this setting, the convergence of evidence approach is fully realized as analysts join together in a committee formed by (agricultural) commodity. Final commodity production estimates are achieved by committee consensus" (NASA, 2004a, p. 4).

The convergence methodology is at the heart of analysis and the final step prior to official world agricultural production estimates and suggests that uncertainty inherent in data and automated models at earlier stages of the analysis are "scrubbed" in a broader context at this final stage.

5. Global change information and PECAD

The relationship between climate and agriculture is complex. Agriculture is not only influenced by a changing climate, but agricultural practices themselves are a contributory factor through emissions of greenhouse gases and influences on fluxes of carbon through photosynthesis and respiration. In short, agriculture is both a contributor to and a recipient of the effects of a changing climate (Rosenzweig, 2003).

At present, PECAD is not directly used to address these dimensions of the climate-agriculture interaction. However, many of the data inputs for PECAD are climate-related, thereby enabling PECAD to inform the understanding of agriculture as a "recipient" of climate-induced changes in temperature, precipitation, soil moisture, and other variables. If reliable climate change prediction of temperature, precipitation, soil moisture, and other necessary variables become available, then these variables can be used as input to PECAD and the

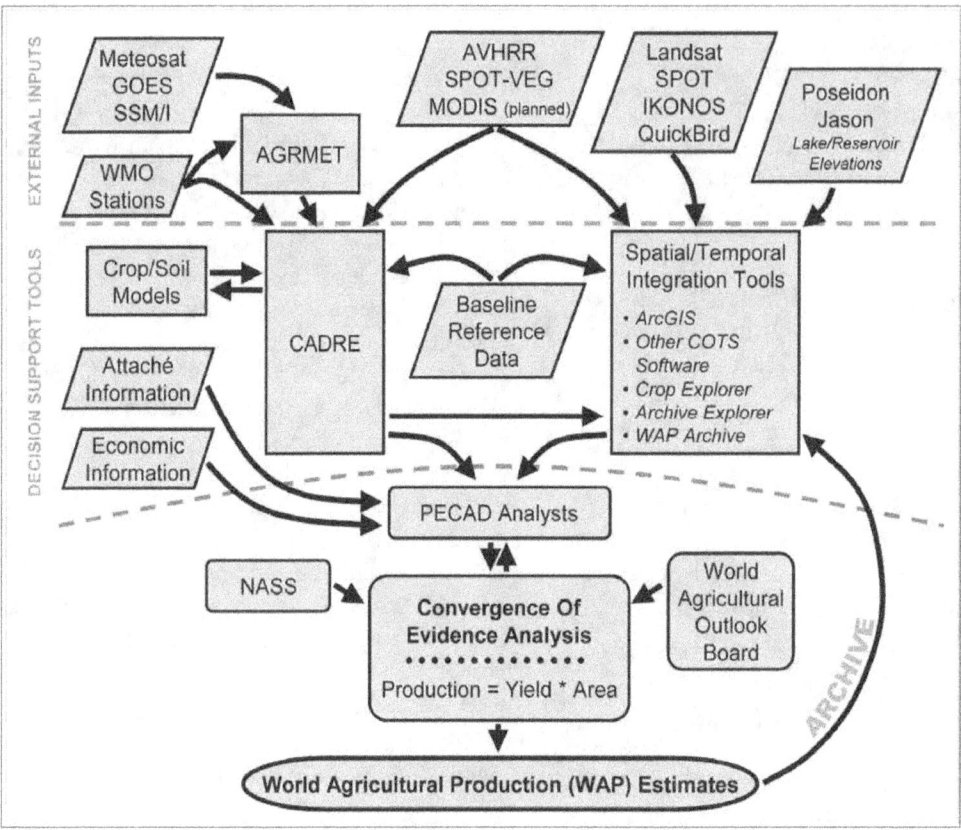

COTS = commercial off-the-shelf; NASS = National Agricultural Statistics Service; WMO = World Meteorological Organization

Figure 1-3 The PECAD DSS: The Role of Convergence of Evidence Analysis (Source: NASA, 2004a, p. 8).

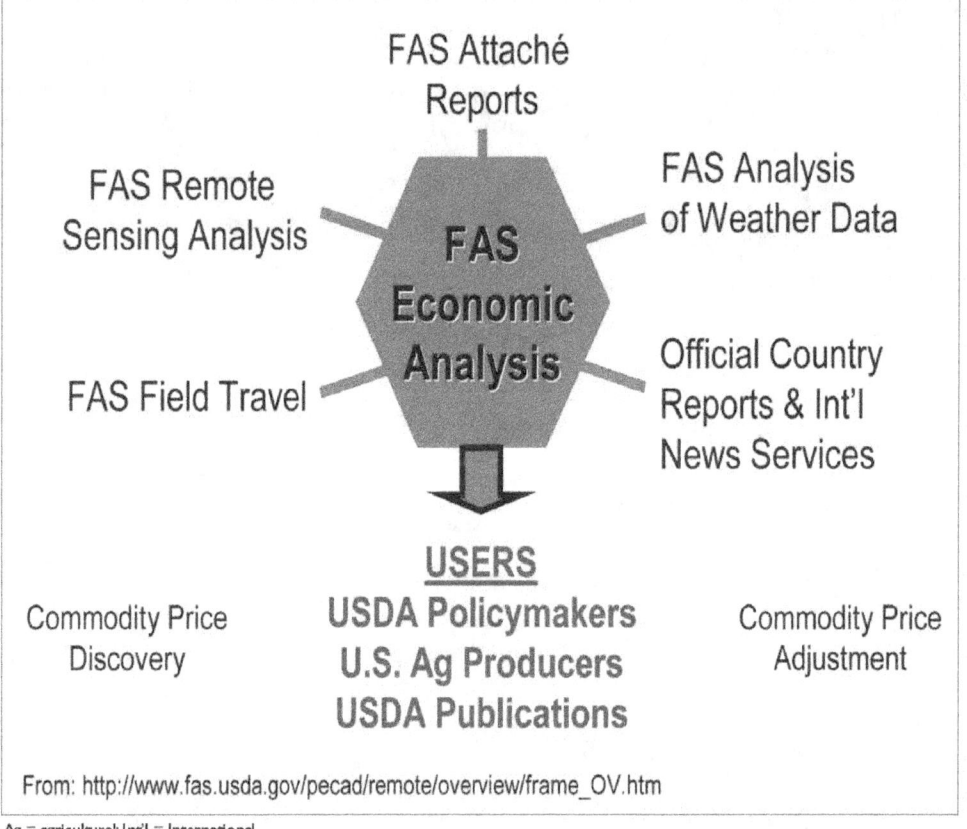

Ag = agricultural; Int'l = International

Figure 1-4 The PECAD DSS: Information Sources for the Convergence of Evidence Analysis (Source: NASA, 2004a, p. 5).

results may be used to provide long-range planning of agricultural practices. In addition, spatial and geographic trends in the output measures from PECAD have the potential to contribute to the understanding of how the agricultural sector is responding to a changing climate.

The output measures of PECAD also can serve to inform the understanding of agriculture as a "contributor" to climate changes. For example, observing trends in PECAD's measures of production and composition of crops can shed light on the contribution of the agriculture sector to agricultural soil carbon sequestration.

The effects of a changing climate on agricultural efficiency as measured by PECAD

PECAD relies on several data sources for agro-meteorological phenomena that affect crop production and the quality of agricultural commodities. These include data that are influenced by climate (e.g., precipitation, temperatures, snow depth, and soil moisture). The productivity measures from PECAD (yield multiplied by area) can also be influenced by climate-induced changes in these data.

In addition, the productivity measures of PECAD can be indirectly but significantly affected by possible climate-induced changes in land use. Examples of such changes include the reallocation of land from food production to biomass fuel production or from food production to forestry cultivation as a means of carbon sequestration. In all of these cases, Earth observations can contribute to understanding climate-related effects on agricultural efficiency (NRC, 2007). Much of the research to integrate Earth observations into climate and agriculture DSTs is relatively recent; for example, in fiscal year 2005, NASA and the USDA began climate simulations using the Goddard Institute for Space Studies (GISS) global climate model (GCM) ocean temperature data and also completed fieldwork for verification and validation of a climate-based crop yield model (NASA, 2006b). The United Nations Food and Agriculture Organization has begun to coordinate similar research on integrating Earth observations and DSSs to study possible effects of changing climate on food production and distribution (e.g., see United Nations Food and Agriculture Organization, no date).

The effects of agricultural practices and efficiency on climate:

The crop assessments and estimates from PECAD, by revealing changes in agricultural practices, could play a role as early indicators to inform forecasting future agricultural-induced effects on climate. The Agricultural Research Service within USDA and NASA have

undertaken research using Earth observation data to study scale-dependent Earth-atmosphere interactions, suggesting that significant changes in regional land use or agricultural practices could affect local and regional climate (NASA, 2001).

2

CHAPTER

Decision Support for Air Quality
Use of CMAQ as a Decision Support Tool for Air Quality to Climate Change

Lead Author: Daewon W. Byun

1. Introduction

Our ability to understand and forecast the quality of the air we breathe, as well as our ability to understand the science of chemical and physical atmospheric interactions, is at the heart of models of air quality. The quality of air is affected by and has implications for the topics presented in our other chapters. Air quality is affected by energy management and agricultural practices, for instance, and is a major factor in public health. Models of air quality also provide a means of evaluating the effectiveness of air pollution and emission control policies and regulations.

While numerous studies examine the potential impact of climate change on forests and vegetation, agriculture, water resources, and human health (examples are found in Brown et al., 2004; Mearns, 2003; Leung and Wigmosta, 1999; Kalkstein and Valimont, 1987), attempts to project the response of air quality to changes in global and regional climates have long been hampered by the absence of proper tools that can transcend the different spatial and temporal scales involved in climate predictions and air quality assessment and by the uncertainties in climate change predictions and associated air quality changes.

One of the popular modeling tools to study air quality as a whole, including tropospheric ozone, fine particles, toxics, acid deposition, and visibility degradation, is the United States (U.S.) Environmental Protection Agency's (EPA) Community Multiscale Air Quality (CMAQ) modeling system. CMAQ's primary objectives are to (1) improve the ability of environmental managers to evaluate the impact of air quality management practices for multiple pollutants at multiple scales, (2) enhance scientific ability to understand and model chemical and physical atmospheric interactions (http://www.epa.gov/

asmdnerl/CMAQ/), and (3) guide the development of air quality regulations and standards and to create state implementation plans. It has been also used to evaluate longer-term pollutant climatology as well as short-term transport from localized sources, and it can be used to perform simulations using downscaled regional climate from global climate change scenarios listed in Intergovernmental Panel on Climate Change (IPCC) (2000). Various observations from the ground and in-situ and from aircrafts and satellite platforms can be used at almost every step of the processing of this Decision-Support Tool (DST) for air quality.

Although there are significant effects of long-range transport, most of the serious air pollution problems are caused by meteorological and chemical processes and their changes at regional and local areas—scales much smaller than those resolved by global climate models (GCM), which are typically applied at a resolution of several hundred kilometers. Current-day regional climate simulations, which typically employ horizontal resolutions of 30 to 60 kilometers (km), are insufficient to resolve small-scale processes that are important for regional air quality, including low-level jets, land-sea breezes, local wind shears, and urban heat island effects (Leung et al., 2006). In addition, climate simulations place enormous demands on computer storage. As a result, most climate simulations only archive a limited set of meteorological variables, the time interval for the archive is usually 6 to 24 hours (e.g., Liang et al., 2006), and some critical information required for air quality modeling is missing.

The interaction and feedback between climate and air chemistry is another issue. Climate and air quality are linked through atmospheric chemical, radiative, and dynamic processes at multiple scales. For instance, aerosols in the atmosphere may modify atmospheric energy fluxes by attenuating, scattering, and absorbing solar and infrared radiation and may also modify cloud formation by altering the growth and droplet size distribution in the clouds. The changes in energy fluxes and cloud fields may, in turn, alter the concentration and distribution of aerosols and other chemical species. Although a few attempts have been made to address these issues, our understanding of climate change is based largely on modeling studies that have neglected these feedback mechanisms.

The impact of climate change on air emissions is also of concern. Changes in temperature, precipitation, soil moisture patterns, and clouds associated with global warming may directly alter emissions, including biogenic emissions (e.g., isoprene and terpenes). Isoprene, an important natural precursor of ozone, is emitted mainly by deciduous tree species. Emission rates are dependent on the availability of solar radiation in the visual range and are highly temperature sensitive. Emissions of terpenes (semi-volatile organic species) may induce formation of secondary organic aerosols. The accompanying changes in the soil moisture, atmospheric stability, and flow patterns complicate these effects, and it is difficult to predict whether climatic change will eventually lead to increased degradation of air quality.

This chapter discusses how CMAQ is used as the DST for studying climate change impact on air quality to address the focus areas required by the Synthesis and Assessment Product (SAP) 5.1 Prospectus: (1) observational capabilities used in the DST, (2) agencies and organizations responsible, (3) characterization of interactions between users and the DST information producers, (4) sources of uncertainties with observation and the decision-support tools, and (5) description of the relation between the DST and climate change information.

2. Description of CMAQ

The U.S. EPA CMAQ modeling system (Byun and Ching, 1999; Byun and Schere, 2006) has the capability to evaluate relationships between emitted precursor species and ozone at urban/regional scales (Appendix W to Part 51 of 40 Code of Federal Regulations: Guideline on Air Quality Models in "http://www.epa.gov/fedrgstr/EPA-AIR/1995/August/Day-09/pr-912.html"). CMAQ uses state-of-the-science techniques for simulating all atmospheric and land processes that affect the transport, transformation, and deposition of atmospheric pollutants. The primary modeling components in the CMAQ modeling system include (1) a meteorological modeling system (e.g., the National Center for Atmospheric Research [NCAR]/Penn State mesoscale model version 5 [MM5]) or a Regional Climate Model (RCM) for the description of atmospheric states and motions, (2) inventories of man-made and natural emissions of precursors that are injected into the atmosphere, and (3) the CMAQ Chemistry Transport Modeling system for

the simulation of the chemical transformation and fate of the emissions. The model can operate on a large range of time scales from minutes to days to weeks as well as on numerous spatial (geographic) scales ranging from local to regional to continental.

The base CMAQ system is maintained by the U.S. EPA. The Center for Environmental Modeling for Policy Development, University of North Carolina at Chapel Hill, is contracted to establish a Community Modeling and Analysis System (CMAS) (http://www.cmascenter.org/) for supporting community-based air quality modeling. CMAS helps development, application, and analysis of environmental models and helps distribution of the DST and related tools to the modeling community. The model performance has been evaluated for various applications (e.g., Zhang et al., 2006; Eder et al., 2006; Tong and Mauzerall, 2006; Yu et al., 2007). Table 2-1 lists the Earth observations (of all types—remote sensing and in situ) presently used in the CMAQ DST.

Within this overall DST structure as shown in Table 2-1, CMAQ is an emission-based, three-dimensional (3-D) air quality model that does not utilize daily observational data directly for the model simulations. The databases utilized in the system represent typical surface conditions and demographic distributions. An example is the EPA's Biogenic Emissions Land Use Database, version 3 (BELD3) database (http://www.epa.gov/ttn/chief/emch/biogenic/) that contains land use and land cover as well as demographic and socioeconomic information. At present, the initial conditions are not specified using observed data even for those species routinely measured as part of the controlled criteria species listed in the National Clean Air Act and its Amendments in an urban area using a dense measurement network. This is because of the difficulty in specifying multi-species conditions that satisfy chemical balance in the system, which is subject to the diurnal evolution of radiative conditions and the atmospheric boundary layer as well as temporal changes in the emissions that reflect constantly changing human activities.

The outputs of the CMAQ and its DST are the concentration and deposition amount of atmospheric trace gases and particulates at the grid resolution of the model, usually at 36 km for the continental U.S. domain and 12 km or 4 km for regional or urban scale domains. The end users of the DST want information on the major scientific uncertainties and our ability to resolve them subject to the information on socioeconomic context and impacts. They seek information on the implications at the national, regional, and local scales and on the baseline and future air quality conditions subject to climate change to assess the effectiveness of current and planned environmental

Table 2-1 Input Data Used for Operating the CMAQ-based DST.

Dataset	Type of Information	Source	Usage
Regional climate model output	Simulation results from an RCM used as a driver for CMAQ modeling; processed through meteorology-chemistry interface processor	RCM modeling team; Pacific Northwest National Laboratory (PNNL), University of Illinois at Urbana Champaign (UIUC), National Center for Environmental Prediction (NCEP), EPA, and universities	Regional climate characterization, driver data for air quality simulations, and emissions processing
Land use, land cover, subsoil category, and topography data; topography for meteorological modeling	Describes land surface conditions and vegetation distribution for surface exchange processes	Various sources from U.S. Geological Survey (USGS), National Astronautics and Space Agency (NASA), NCEP EPA, states, etc.	Usually the data are associated with RCMs land surface module; need to be consistent with vegetation information, such as BELD3 if possible
Biogenic emissions land use database version 3 (BELD3)	Land use and biomass data and vegetation/tree species fractions	EPA	Processing of biogenic emissions; used to provide activity data for county-based emission estimates; now also used for land surface modeling in RCM
Air emissions inventories: national emissions inventories and state/special inventories; often called "bottom-up" inventories	Amount and type of pollutants into the atmosphere: Chemical or physical identity of pollutants Geographic area covered Institutional entities Time period over which the emissions are estimated Types of activities that cause emissions	EPA, regional program organizations, state and local government, and foreign governments	Preparation of model-ready emission inputs; perform speciation for the chemical mechanism used; used to evaluate "top-down" emissions (i.e., from inversion of satellite observations though air chemistry models)
Chemical species initial and boundary conditions	Clean species concentration profiles initial input and boundary conditions used for CMAQ simulations; originally from observations from clean background locations	EPA (fixed profiles), Goddard Earth Observing System (GEOS)-Chemical (GEOS-Chem) (Harvard and University of Houston), Model of Ozone and Related Chemical Tracers (MOZART) (NCAR); dynamic concentrations with diurnal variations (daily, monthly or seasonal)	CMAQ simulations; fixed profiles are used for outer domains where no significant emissions sources are located
Archived databases: Air Quality System (AQS)/ AIRNow	Near real-time (AIRNow) and AQSs for ozone, particulate matter, and some toxic species	Joint partnership between EPA and state and local air quality agencies	Measurement data used for model evaluations; report and communicate national air quality conditions for emissions control decision support

policies. Local air quality managers would want to know if the DST could help assess methods of attaining current and future ambient air quality standards and evaluate opportunities to mitigate the climate change impacts. Decision makers would ask modelers to simulate the air quality in the future for a few plausible variations in the model inputs that represent plausible climate scenarios of regional implications. Through sensitivity simulations of the DST with different assumptions on the meteorological

and emissions inputs, the effectiveness of such policies and uncertainties in the system can be studied. The results can be also compared with the historic air quality observations with similar ambient conditions to validate predictions of the DST.

3. Potential Future Uses and Limits

Although one of the major strengths of CMAQ is
its reliance on the first principles of physics and
chemistry, a few modeling components, such as cloud
processes, fine scale turbulence, radiative processes,
etc., rely on parameterizations or phenomenological
concepts to represent intricate and less well known
atmospheric processes. The present limitations in
science parameterizations and modeling difficulties will
continuously be improved as new understanding of these
phenomena are obtained through various measurements
and model evaluation/verification. The development
of the chemical mechanism, Carbon Bond 05, which
recently replaced Carbon Bond 04 is a case in point.
The reliability of the CMAQ simulation result is subject
to quality of the emission inputs, both at the global and
regional scales, which depend heavily on socioeconomic
conditions. Because such estimates are obtained using
projection models in relevant socioeconomic disciplinary
areas, their accuracy must be scrutinized when used for
the decision-making process. The CMAQ DST users/
operators may not always have domain expertise to
discern the validity of such results.

CMAQ needs to have the capability to utilize available
observations to specify more accurately the critical
model inputs, although they have been chosen based on
best available information and current experience. A data
assimilation approach may be used to improve the system
performance at different processing steps.

For example, research has been undertaken to use
satellite remote-sensing data products together with
high-resolution land use and land cover data to improve
the land-surface parameterizations and boundary layer
schemes in the RCMs (e.g., Pour-Biazar et al., 2007).
Active research in chemical data assimilation (e.g.,
Constantinescu et al., 2007a and b) is currently conducted
with models such as Sulfur Transport Eulerian Model
(STEM)-II (Carmichael et al., 1991) and GEOS-Chem
(Bey et al., 2001) , which utilize both in-situ and satellite
observations (e.g., Sandu et al., 2005; Kopacz et al.,
2007; Fu et al., 2007). Because of the coarse spatial and
temporal resolutions of the satellite data collected in the
1960s through the 1980s and gas measurements through
the launch of Earth Observing System (EOS) Aura in
2004, most research in this area has been performed with
global chemistry-transport models. As the horizontal
footprints of modern satellite instruments reach the
resolution suitable for regional air quality modeling,
these data can be used to evaluate and then improve the
bottom-up emissions inputs in the regional air quality
models. However, they do not provide required vertical
information. The exception is occultation instruments,

but these do not measure low enough in altitude for
near-surface air quality applications. In-situ and remote-
sensing measurements from ground and aircraft platforms
could be used to augment the satellite data in these data
assimilation experiments.

Utilization of the column-integrated satellite
measurements in a high-resolution, 3-D grid model
like CMAQ poses serious challenges in distributing the
pollutants vertically and separating those within and
above the atmospheric boundary layer. Because similar
problems exist for the retrieval of meteorological profiles
of moisture and temperature, experiences that include
these can be adapted for a few well-behaved chemical
species. A data assimilation tool can be used to improve
the initial and boundary conditions using various in-situ
and satellite measurements of atmospheric constituents.
At present, however, an operational assimilation system
for CMAQ is not yet available, although prototype
assimilation codes have recently been generated (Hakami
et al., 2007; Zhang et al., 2007). Should these data
assimilation tools become part of the DST, various
conventional and new satellite products, including
Tropospheric Emission Spectrometer ozone profiles,
Geostationary Operational Environmental Satellites
(GOES) hourly total ozone column (GhTOC) data, Ozone
Monitoring Instrument (OMI) total ozone column (TOC),
the Cloud-Aerosol Lidar and Infrared Pathfinder Satellite
Observation (CALIPSO) (http://www-calipso.larc.nasa.
gov/) attenuated backscatter profiles, and OMI aerosol
optical thickness (AOT) data, can be utilized to improve
the urban-to-regional scale air quality predictions.

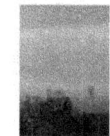

Because of the critical role of the RCM as the driver of
CMAQ in climate change studies, RCM results for the
long-term simulations must be verified thoroughly. To
date, evaluation of the RCM has been performed for
the air quality-related operations only for relatively
short simulation periods. For example, the simulated
surface temperature, pressure, and wind speed must be
compared to surface observations to determine how well
the model captures the mean land-ocean temperature and
pressure gradients, the mean sea breeze wind speeds,
the average inland penetration of sea breeze, the urban
heat island effect, and the seasonal variations of these
features. Comparisons with rawinsonde soundings and
atmospheric profiler data would determine how well the
model reproduces the averaged characteristics of the
afternoon mixed layer heights and of the early morning
temperature inversion as well as the speed and the vertical
wind shears of the low-level jets. In addition to these
mesoscale phenomena, changes in other factors can also
alter the air pollution patterns in the future and need to
be carefully examined. These factors include the diurnal
maximum, minimum, and mean temperature; cloud

cover; thunderstorm frequency; surface precipitation and soil moisture patterns; and boundary layer growth and nocturnal inversion strength.

In global model applications, it has been demonstrated that satellite measured biomass burning emissions data are necessary to enhance model predictability (e.g., Duncan et al., 2003; Hoelzemann et al., 2004). Duncan et al. (2003) presented a methodology for estimating the seasonal and interannual variation of biomass burning, which was designed for use in global chemical transport models using fire-count data from the Along Track Scanning Radiometer and the Advanced Very High Resolution Radiometer (AVHRR) World Fire Atlases. The Total Ozone Mapping Spectrometer (TOMS) Aerosol Index data product was used as a surrogate to estimate interannual variability in biomass burning. Also Spracklen et al. (2007) showed that the wildfire contribution to the interannual variability of organic carbon aerosol can be studied using the area-burned data and ecosystem specific fuel loading data. A similar fire emissions dataset at the regional scales could be developed for use in a study of climate impact on air quality. For retrospective application, a method similar to that used by the National Oceanic and Atmospheric Administration's (NOAA) Hazard Mapping System for Fire and Smoke (http://www.ssd.noaa.gov/PS/FIRE/hms.html) may be used to produce a long-term regional scale fire emissions inventory for climate impact analysis.

4. Uncertainty

The CMAQ modeling system as currently operated has several sources of uncertainty in addition to those associated with some of the limits described in the previous section. In particular, when CMAQ is used to study the effects of climate change and air quality, improvements in several areas are necessary to reduce uncertainty. First, the regional air quality models employ limited modeling domains and, as such, they are ignorant of air pollution events outside the domains unless proper dynamic boundary conditions are provided. Second, because the pollutant transport and chemical reactions are fundamentally affected by the meteorological conditions, improvement of both the global and regional climate models and the downscaling methods by evaluating and verifying physical algorithms that have been implemented with observations is necessary to improve the system's overall performance. Third, the basic model inputs, including land use/vegetation cover descriptions and emissions inputs must be improved. Fourth, the model representativeness issues, including grid resolution problems, compensating errors among the model components, and incommensurability of the model results compared with the dimensionality of the measurements (i.e., inherent differences in the modeled outputs that

represent volume and time averaged quantities to the point or path-integrated measurements) as discussed in Russell and Dennis (2000) and NARSTO (2000), need to be addressed. These factors are the principal cause of simulation/prediction errors.

Although the models incorporated in this DST are first principle-based environmental models, they have difficulties in representing forcing terms in the system, particularly, the influence of the earth's surface, long-range transport, and uncertainties in the model inputs such as daily emissions changes due to anthropogenic and natural events. There is ample opportunity to reduce some uncertainties associated with CMAQ through model evaluation and verification using current and future meteorological and atmospheric chemistry observations. Satellite data products assimilated in the global chemical transport models (GCTM) could provide better dynamic lateral boundary conditions for the regional air quality modeling (e.g., Al-Saddi et al., 2005). Additional opportunities to reduce the model uncertainty include comparison of model results with observed data at different resolutions, quantification of effects of initial and boundary conditions and chemical mechanisms, application of CMAQ to estimate the uncertainty of input emissions data, and ensemble modeling (using a large pool of simulations among a variety of models) as a means to estimate model uncertainty.

A limitation in CMAQ applications, and therefore a source of uncertainty, has been the establishment of initial conditions. The default initial conditions and lateral boundary conditions in CMAQ are provided under the assumption that after spin-up of the model, they no longer play a role, and in time, surface emissions govern the air quality found in the lower troposphere. Song et al. (2007) showed that the effects of the lateral boundary conditions differ for different latitudes and altitudes as well as seasons. In the future, dynamic boundary conditions can be provided by fully integrating the GCTMs as part of the system. Several research groups are actively working on this, but the simulation results are not yet available in open literature. A scientific cooperative forum, the Task Force on Hemispheric Transport of Air Pollution (http://www.htap.org/index.htm), is endeavoring to bring together the national and international research efforts at the regional, hemispheric, and global scales to develop a better understanding of air pollution transport in the Northern Hemisphere. This task force is currently preparing its 2007 Interim Report addressing various long-range transport of air pollutant issues (http://www.htap.org/activities/2007_Interim_Report.htm). Although the effort does not directly address climate change issues, many of findings and tools used are very relevant to meteorological and chemical downscaling issues.

Ultimately, CMAQ should consider all the uncertainties in the inputs. The system's response may be directly related to the model configuration and algorithms (e.g., structures, resolutions, and chemical and transport algorithms), compensating errors, and the incommensurability of modeling nature as suggested by Russell and Dennis (2000).

5. Global Change Information and CMAQ

CMAQ could be used to help answer several questions about the relationship between air quality and climate change, including the following:

1. How will global warming affect air quality in a region?

2. How will land use change due to climate change and urbanization, or how will intentional management decisions affect air quality?

3. How much will climate change alter the frequency, seasonal distribution, and intensity of synoptic weather patterns that influence pollution in a region?

4. How sensitive are air quality simulations to uncertainty in wildfire projections and to potential land management scenarios?

5. How might the contribution of the local production and long-range transport of pollutants differ due to different climate change scenarios?

6. Will future emissions scenarios or climate changes affect the frequency and magnitude of high pollution events?

To provide answers to these questions, CMAQ will rely heavily on climate change-related information. In addition to the influence of greenhouse gases and global warming, other forcing functions include population growth, land use changes, new emission controls being implemented, and the availability of new energy sources to replace the existing high-carbon sources. Different scenarios can be chosen either to study potential impacts or to estimate the range of uncertainties of the predictions. The two upstream climate models, GCMs and RCMs, generate the climate change data that drive a GCTM and CMAQ. Both the GCMs and RCMs are expected to represent future climate change conditions while simulating historic climate conditions that can be verified with comprehensive datasets such as the NCEP Reanalysis data provided by the NOAA/Oceanic and Atmospheric Research/Earth Systems Research Laboratory (ESRL) Physical Sciences Division, Boulder, Colorado, from their Web site, http://www.cdc.noaa. gov/cdc/data.ncep.reanalysis.html. The meteorology simulated by the climate models represents conditions in future year scenarios, reflecting changing atmospheric conditions. Furthermore, emissions inputs used for the GCTM and CMAQ must reflect the natural changes and/or anthropogenic developments related to climate change and other factors (e.g., population growth and geographical population shifts due to climate change).

In recent years, the EPA Science to Achieve Results (STAR) program has funded several projects on the possible effects of climate change on air quality and ecosystems. A majority of these projects have adopted CMAQ as the base study tool. Figure 2-1 provides a general schematic of the potential structure of a CMAQ-based climate change DST. The figure shows potential uses of CMAQ for climate study; most climate-related CMAQ applications are not yet configured as fully as indicated in the figure.

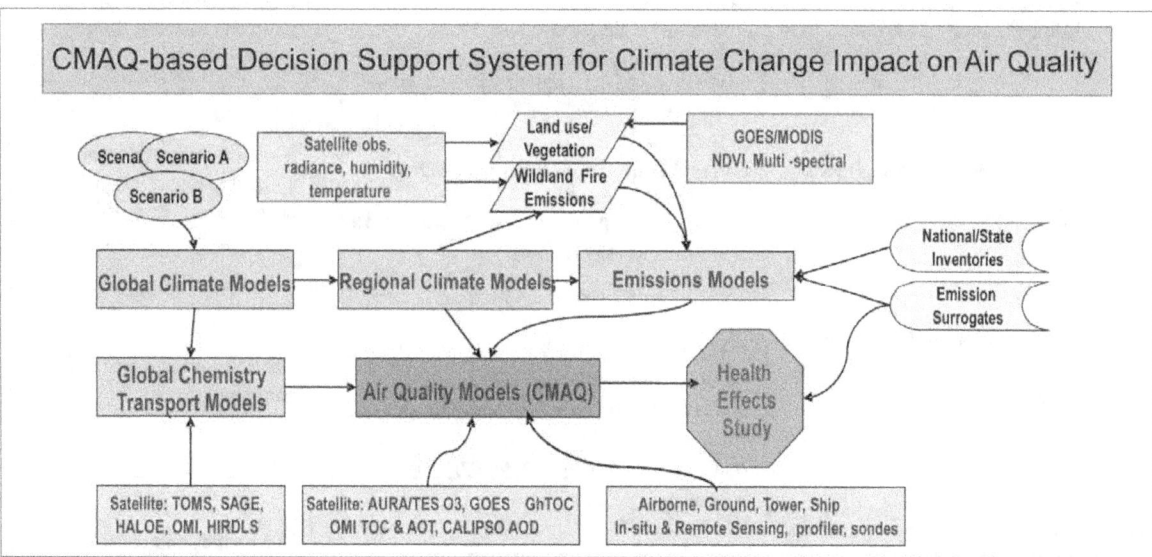

AOD = aerosol optical depth; HALOE = Halogen Occultation Experiment; HRDLS = High Resolution Dynamics Limb Sounder ; SAGE = Stratospheric Aerosol and Gas Experiment; TES = Tropospheric Emission Spectrometer

Figure 2-1 Configuration of CMAQ-based DST for Climate Change Impact Study

The projects linking CMAQ and climate study have used upstream models and downstream tools, including those identified in Table 2-1. Related projects that use regional air quality models other than CMAQ are also listed. For the GCMs, the NCAR Community Climate Model (CCM) (Kiehl et al., 1996), NASA Goddard Institute for Space Studies (GISS) model (e.g., Hansen et al., 2005), and NOAA Geophysical Fluid Dynamics Laboratory (GFDL) Climate Model 2 (CM2) (Delworth et al., 2006) are the most popular global models for providing meteorological inputs representing climate change events. A recent description for the GISS model can be found in Schmidt et al. (2006) (http://www.giss.nasa.gov/tools/) and for the CCM in Kiehl et al. (1996) (http://www. cgd.ucar.edu/cms/ccm3/). A newer version of the CCM was released on May 17, 2002 with a new name—the Community Atmosphere Model (http://www.ccsm.ucar.

edu/models/atm-cam). The model is described in Hurrell et al. (2006).

As shown in Table 2-2, for climate change studies, CMAQ is linked with upstream models such as a GCM, a global tropospheric chemistry model (GTCM), and an RCM to provide emissions sensitivity analysis, source apportionment, and data assimilation to assist policy and management decision-making activities, including health impact analysis. Certain EPA STAR projects (Hogrefe et al., 2004 and 2005; Knowlton et al., 2004; Civerolo et al., 2007) have utilized the CMAQ-based DST to assess whether climate change would influence the effectiveness of current and future air pollution policy decisions subject to the potential changes in local and regional meteorological conditions.

Table 2-2 Illustrative Example of the Potential Uses of the Models and Upstream and Downstream Tools for a CMAQ-based Climate Change Impact DST

Component	Functions	Model Name: Owner	Users
GCM	Performs climate change simulations over the globe for different Special Report on Emissions Scenarios (SRES) climate scenarios; typical resolution for a long-term (50 year) simulation is at 4° x 5° latitude and longitude	CCM: NCAR Goddard Institute for Space Studies (GISS) GCM: NASA CM2: GFDL of NOAA	Climate research institutes, universities, and government institutions
GCTM	Computes global scale chemical states in the atmosphere; uses same resolution as GCM	GEOS-Chem: NASA, Harvard University MOZART: NCAR (Earth and Sun Systems Laboratory [ESSL]/Atmospheric Chemistry Division)	Global chemistry research organizations, universities, and government institutions
RCM	Simulates regional scale climate and meteorological conditions downscaling the GCM output; for U.S. application ~36-km resolution used	MM5-based: NCAR, PNNL, UIUC, and others The weather research and forecasting (WRF)/advanced research WRF (ARW) core based: NCAR, UIUC Eta-based: NCEP (before June 2006) The WRF-nonhydrostatic mesoscale model (NMM) core based: NCEP (after June 2006)	Regional climate research groups, universities, and government institutions
Regional air quality models	Performs air quality simulations at regional and urban scales at the same resolution as the RCM	CMAQ: EPA Comprehensive air quality model with extensions: Environment WRF-Chemical (WRF-Chem): NOAA/NCAR STEM-II: University of Iowa	Regional, state, and local air quality organizations; universities; private industries; and consulting companies
Downstream tools for decision support	Performs additional computations to help decision support, such as sensitivity, source apportionment, and exposure studies	CMAQ/ Decoupled Direct Method: Georgia Institute of Technology CMAQ/4-dimensional variable: CalTech/Virginia Tech/University of Houston Stochastic human exposure and dose simulation: EPA Total risk integrated methodology: EPA	Universities and consulting companies
Upstream tools for representing climate change impacts on input data	Performs additional computations to generate model inputs that affect simulations	Land-surface models SLEUTH: USGS and University of California, Santa Barbara (captures urban patterns) Community land model: NCAR (used for RCM and biogenic emission estimates after growth)	Universities and consulting companies

Other EPA STAR projects employ global climate change information from a GCM. For example, Tagaris et al. (2007) and Liao et al. (2007) use the results of GCM simulation with the well-mixed greenhouse gases—CO_2, CH_4, N_2O, and halocarbons—updated yearly from observations for 1950 to 2000 (Hansen et al., 2002) and for 2000 to 2052 following the A1B SRES scenario from the Intergovernmental Panel on Climate Change (IPCC) 2001. The simulation used ozone and aerosol concentrations in the radiative scheme fixed at present-day climatological value provided in Mickley et al. (2004).

To resolve the meteorological features affecting air pollution transport and transformation at a regional scale, the coarse scale meteorological data representing the climate change effects derived from a GCM are downscaled using an RCM. An RCM is often based on a limited-domain, regional mesoscale model (such as mesoscale model version 5 [MM5]), the Regional Atmospheric Modeling System, Eta, and the WRF-ARW or WRF-NMM. An alternative method for constructing regional scale climate change data is through statistical downscaling, which evaluates observed spatial and temporal relationships between large-scale (predictors) and local (predictands) climate variables over a specified training period and domain (Spak et al., 2007). Because of the need to use a meteorological driver that satisfies constraints of dynamic consistency (i.e., mass and momentum conservations) for regional scale air quality modeling (e.g., Byun, 1999 a and b), the CMAQ modeling system relies exclusively on the dynamic downscaling method.

Regional chemistry/transport models, like CMAQ, are better suited for regional air quality simulations than a GCTM because of the acute air pollution problems that are managed and controlled through policy decisions at specific geographic locations. Difficulty in prescribing proper boundary conditions, especially in the upper troposphere, is one of the deficiencies of CMAQ simulations of air quality (e.g., Tarasick et al., 2007; Tang et al., 2007). Therefore, one of the main roles of the GCTM is to provide proper dynamic boundary conditions for CMAQ to represent temporal variation of chemical conditions that might be affected by the long-range transport of pollution (e.g., particle from large-scale biomass burnings) from outside the regional domain boundaries (Holloway et al., 2002; In et al., 2007). The contemporary EPA-funded projects on climate change impact on air quality mainly use two 3-D GCTM models: the NASA/Harvard GEOS-Chem (Bey et al., 2001) and the NCAR MOZART (Brasseur et al., 1998; Horowitz et al., 2003).

The GEOS-Chem model (http://www-as.harvard.edu/chemistry/trop) is a global model for predicting tropospheric composition. The model was originally driven by the assimilated meteorological observation data from the GEOS of the NASA Global Modeling and Assimilation Office (GMAO). GEOS-Chem has been used as community assessment models for NASA Global Model Initiative, climate change studies with the NASA/GISS GCM, chemical data assimilation of tropospheric gaseous and aerosol species at NASA GMAO, and regulatory models for air pollution, particularly in providing long-range transport information for regional air quality models. Long-term retrospective studies are possible with GEOS data, which are available from 1985 to present at a horizontal resolution of 2 degrees (latitude) by 2.5 degrees (longitude) until the end of 1999 and 1 degree by 1 degree afterward. For climate studies, the NASA GISS GCM meteorological outputs are used instead. Emission inventories include a satellite-based inventory of fire emissions (Duncan et al., 2003) with expanded capability for daily temporal resolution (Heald et al., 2003) and the National Emissions Inventory for 1999 for the U.S. with monthly updates in order to achieve adequate consistency with the CMAQ fields at the GEOS-Chem/CMAQ interface.

MOZART (http://gctm.acd.ucar.edu/mozart/models/m3/index.shtml) is built on the framework of the Model of Atmospheric Transport and Chemistry that can be driven with various meteorological inputs and at different resolutions, such as meteorological reanalysis data from NCEP, NASA GMAO, and the European Centre for Medium-Range Weather Forecasts. For climate change applications, meteorological inputs from the NCAR Community Climate Model (CCM3) are used. MOZART includes a detailed chemistry scheme for tropospheric ozone, nitrogen oxides, and hydrocarbon chemistry; a semi-Lagrangian transport scheme; dry and wet removal processes; and emissions inputs. Emission inputs include sources from fossil fuel combustion, biofuel and biomass burning, biogenic and soil emissions, and oceanic emissions. The surface emissions of NOx, CO, and NMHCs are based on the inventories described in Horowitz et al. (2003), aircraft emissions based on Friedl (1997), and lightning NOx emissions that are distributed at the location of convective clouds.

GCTMs are applied to investigate numerous tropospheric chemistry issues involving gases—CO, CH_4, OH, NOx, HCHO, and isoprene—and inorganic (sulfates and nitrates) and organic (elemental and organic carbons) particulates. Various in-situ, aircraft, and satellite-based measurements are used to provide the necessary inputs to verify the science process algorithms and to perform general model evaluations. They include vertical profiles from aircraft observations as compiled by Emmons et al. (2000), multiyear analysis of ozonesonde data (Logan, 1999), and those available at the Community Data Web site managed by the NCAR ESSL Atmospheric Chemistry Division as well as multiyear surface observations of CO reanalysis (Novelli et al., 2003). Current and previous atmospheric measurement campaigns are listed in Web pages by NOAA ESRL (http://www.esrl.noaa.gov/); NASA, Tropospheric Integrated Chemistry Data Center (http://www-air.larc.nasa.gov/); and NCAR ESSL Atmospheric Chemistry Division Community Data (http://www.acd.ucar.edu/Data/). These observations are used to set boundary conditions for the slow reacting species, including CH_4, N_2O, and chlorofluorocarbons (CFC), and to evaluate other modeled species, including CO, NO_x, peroxyacetyl nitrate (PAN), HNO_3, HCHO, acetone, H_2O_2, and non-methane hydrocarbons. In addition, several satellite measurements of CO, NO_2, and HCHO from the Global Ozone Monitoring Experiment, the Scanning Imaging Absorption Spectrometer for Atmospheric Chartography, and OMI instruments have been used extensively to verify the emissions inputs and performance of the GCTM.

The grid resolutions used in the studies discussed above are much coarser than those used in the air quality models for studying emission control policy issues, such as evaluating state implementation plans. State implementation plan modeling typically utilizes over 20 vertical layers with a 4-km horizontal grid spacing to reduce uncertainties in the model predictions near the ground and around high-emission source areas, including urban and industrial centers. Although Civerolo et al. (2007) applied CMAQ at a higher resolution, the duration of the CMAQ simulation was far too short a time scale to evaluate the regional climate impacts in detail.

One of the additional key limitations of using the CMAQ for climate change studies is that the linkages between climate and air quality and from the global scale to regional scale models are only one way (i.e., no feedback). Jacob and Gilliland (2005) stated that one-way assessment of the global change scenarios would be less useful for projection of air pollutant emissions because the evolution of regional air quality policies were not accounted for in these storylines. Also, to represent the interactions between atmospheric chemistry and meteorology, such as radiation and cloud/precipitation microphysics, particulates, and heterogeneous chemistry, a two-way linkage must be established between the meteorology and chemistry models. An online modeling approach as implemented in WRF-Chemistry (WRF-Chem) is an example of such a linkage, but still there is a need to develop a link between the global and regional scales. A multi-resolution modeling system,

such as demonstrated by Jacobson (2001 a, b), might be necessary to address the true linkage between air pollution forcing and climate change and to provide the urban-to-global connection.

In addition, there would be significant benefits to linking other multimedia models describing subsoil conditions, vegetation dynamics, hydrological processes, and ocean dynamics, including the physical/chemical interactions between the ocean micro-sublayer and atmospheric boundary layer to an air quality model. To generate such a mega model under one computer coding structure would require handling of extremely different state variables in each multimedia model with substantially different data. Furthermore, interactions among the multimedia models need multidirectional data inputs, quality assurance checkpoints, and decision-support entries. A more generalized online and two-way data exchange tool currently being developed under the Earth System Modeling Framework (http://www.esmf.ucar.edu/) may be a viable option.

Observations not only represent the real changes in the climate but also provide a fundamental database to verify various modeling components in the DST. The meteorological reanalysis data are available both in regional and global scales, but a similar atmospheric chemistry database for air quality is lacking. An ozone database from the ozonesonde system and other in-situ measurements are useful for global-scale studies. But for regional air quality studies, the availability of such measurements representing long-term urban and local conditions is limited. Satellite or other remote-sensing platform observations may provide additional data sources to build an atmospheric chemistry reanalysis database at global and regional scales, but theses observations are mainly limited to ozone and aerosols. Such a chemical reanalysis database can be utilized to study long-term air quality trends; evaluate science process components in the air quality models, emissions, and other model inputs and configurations; and improve model predictions through data assimilation approaches.

CHAPTER 3

Decision Support System for Assessing Hybrid Renewable Energy Systems

Lead Author: David S. Renne'

1. Introduction

The national application area addressed in this chapter is the deployment of renewable energy technologies. Renewable energy technologies are being used around the world to meet local energy loads, supplement grid-wind electricity supply, perform mechanical work such as water pumping, provide fuels for transportation, provide hot water for buildings, and to support heating and cooling requirements for building energy design. Numerous organizations and research institutions around the world have developed a variety of decision-support tools (DST) to address how these technologies might perform in the most cost-effective manner to address specific applications. This chapter will focus on one specific tool, the Micropower Optimization Model known as the Hybrid Optimization Model for Electric Renewables (HOMER)®*, which has been under consistent development and improvement at the U.S. Department of Energy's National Renewable Energy Laboratory (NREL) and is used extensively around the world (Lambert et al., 2006).

HOMER relies heavily on knowledge of the renewable energy resources available to the technologies being analyzed. Renewable energy resources, particularly for solar and wind technologies, are highly dependent on weather and climate phenomena and are also driven by local microclimatic processes. Given the absence of a sufficiently dense ground network of reliable solar and wind observations, we must rely on validated numerical models, empirical knowledge of microscale weather characteristics, and collateral (indirect) observations derived from Earth observations (such as reanalysis data and satellite-borne remote sensors) to develop reliable knowledge of the geospatial characteristics and extent of these resources. Thus, the Decision-Support System

(DSS) described in this chapter includes HOMER as an end-use application and is described in the context of the renewable energy resource information required as input, as well as some intermediate steps that can be taken to organize these data, using Geographic Information Systems (GIS) software to facilitate the application of HOMER.

2. Description of the HOMER DSS

The HOMER DSS described in this chapter consists of three main components: (1) the renewable energy resource information required to estimate technology performance and operational characteristics, (2) (optional) organization of the resource data into a GIS framework so that the data can be easily imported into the DST, and (3) NREL's Micropower Optimization Model known as HOMER, which ingests the renewable resource data for determining the optimal mix of power technologies for meeting specified load conditions at specified locations. This section describes each of these components separately. Although climate-based Earth observational data are primarily relevant only to the first component, some related Earth observation information could also be associated with the second and even the third component. Furthermore, it will be apparent that the first component is of major importance in the successful use of the HOMER DSS.

Although HOMER handles a number of power technologies, we will focus our attention in this chapter on solar and wind technologies and the resources required to run these technologies.

Solar and Wind Resource Assessments

The first component of the HOMER DSS is properly formatted, reliable and renewable energy resource data. The significant data requirements for this component are time-dependent measurements of wind and solar resources as well as Earth observational data and data from numerical models to provide the necessary spatial information for these resources, which can vary significantly over relatively small distances due

* HOMER is a Registered Trademark of Midwest Research Institute, the management and operating contractor of the National Renewable Energy Laboratory (NREL) for the U.S. Department of Energy.

to local microclimatic effects. Because of this natural variability, it is necessary to examine these energy resources geospatially in order to determine optimal siting of renewable energy technologies; alternatively, if a renewable energy technology is sited at a specific site in order to meet a nearby load requirement (such as a solar home system), it is necessary to know what the resource availability is at that location since microclimatic variability may make even nearby data sources irrelevant.

Examples of the products derived from the methodologies described below can be found for many areas around the world. One significant project that has recently been completed is the Solar and Wind Energy Resource Assessment (SWERA) Project, which provided high-resolution wind and solar resource maps for 13 countries around the world. SWERA was a project funded by the Global Environment Facility and was cost-shared by several technical organizations around the world: NREL; the State University of New York at Albany, NASA's Langley Research Center, and the U.S. Geological Survey (USGS)/Earth Resources Observation Systems (EROS) Data Center in the U.S.; Riso National Laboratory in Denmark; the German Aerospace Institute (DLR); the Energy Resources Institute (New Delhi, India); and the Brazilian Spatial Institute in Sao Jose dos Campos, Brazil. The United Nations environment programmer managed the project. Besides the solar and wind resource maps and underlying datasets, a variety of other relevant data products came out of this program. All of the final products and data can be found on the SWERA archive, hosted at the United Nations environment programmer/ Global Resource Information Database site, collocated with the USGS/EROS data center in Sioux Falls, South Dakota (http://swera.unep.net).

For wind resource assessments, NREL's approach, known as the Wind Resource Assessment Mapping System (WRAMS), relies on mesoscale numerical models such as mesoscale model version 5 (MM5) or weather research and forecasting (WRF), which can provide simulations of near-surface wind flow characteristics in complex terrain or where sharp temperature gradients might exist (such as land-sea contrasts). Typically, these numerical models use available weather data, such as the National Climatic Data Center's Integrated Surface Hourly (ISH) data network and National Center for Atmospheric Research (NCAR)-National Centers for Environmental Prediction (NCEP) reanalysis data as inputs. In coastal areas or island situations, NREL's wind resource mapping also relies heavily on SeaWinds data from the Quick Scatterometer (Quickscat) satellite to obtain near-shore and near-island wind resources. WRAMS also relies on Global Land Cover Characterization 1-kilometer (km) and Regional Gap Analysis Program 200-meter (m)

land cover data as well as Moderate Resolution Imaging Spectroradiometer (MODIS) data from the Aqua and Terra Earth Observation System satellites to obtain information such as percent of tree cover and other land use information. This information is used not only to determine roughness lengths in the numerical mesoscale models but also to screen sites suitable for both wind and solar development in the second component of the HOMER DSS.

The numerical models are typically run at a 2.5-km resolution. However, wind resource information is often reported at the highest resolution at which a digital elevation model (DEM) can provide. Globally, this has traditionally been a 1-km resolution; however, in some cases in the U.S., 400-m DEM data are available. Furthermore, the Shuttle Radar Topology Mission (SRTM) has now been able to provide users with a 90-m DEM for much of the world. Thus, additional steps are needed beyond the 2.5-km resolution model output to depict wind resources at the higher resolutions offered by these DEMs. This can be accomplished by using a secondary high-resolution mesoscale model, empirical methods, or both. For example, with NREL's WRAMS methodology, GIS-based empirical modeling tools have been developed to modify results from the numerical models that appear to have provided unreliable results in complex-terrain areas.

The numerical models generally provide outputs at multiple levels above the ground. The WRAMS methodology provides values at a single, specified height above the ground, nominally 50 m, or near the hub-height of modern-day large wind turbines (although with the recent advent of larger and larger wind turbines, hub heights are approaching 100 m, so this standard height designation is changing). Where measured data are used to assess wind resources, a simple "power law" relationship is used to extrapolate the measured data to the desired height (Elliott et al., 1987) as follows:

$$V_R/V_a = (Z_R/Z_a)^\alpha \qquad \text{(1)}$$

where α, the power law coefficient, is normally assumed to be $1/7$, V_R is the wind speed at reference height Z_R (nominally, 50 m), and V_a is the wind speed at the measurement height Z_a.

The output of the WRAMS methodology is typically a value of wind power density at every grid-cell representative of an annual average (in order to produce monthly values, the procedure outlined above would have to be repeated for each month of the year). For mapping purposes, a classification scheme has been set up that relates a "wind power class" to a range of wind power densities. The classification scheme ranges from 1 to >7, and applies to a specific height above ground. Normally for grid-connected applications, a wind power class of 4 or above is best, while for small wind turbine applications where machines can operate in lower wind speeds, a wind power class of 3 or above is suitable. Of course, the wind maps are not intended to identify sites at which large wind turbines can be installed but rather are intended to provide information to developers on where they might most effectively install wind measurement systems for further site assessment. The maps also provide a useful tool for policy makers to obtain reliable estimates on the total wind energy potential for a region.

Other well-known approaches besides NREL's WRAMS methodology are also used to produce large-area wind resource mapping. For example, Riso National Laboratory calculates wind speeds within 200 m above the Earth's surface using the Karlsruhe Atmospheric Mesoscale Model (KAMM). Although KAMM also uses NCEP/NCAR reanalysis data, the model is based on large-scale geostrophic winds, and simulations are performed for classes of different geostrophic wind. The classes are weighted with their frequency to obtain statistics for the simulated winds. The results can then be treated as similar to real observations to make wind atlas files for the Wind Atlas Analysis and Application Program (WAsP), which are employed to predict local winds at a much higher resolution than KAMM can provide. WAsP calculations are based on measured or simulated wind data at specific locations and include a complex terrain flow model, a roughness change model, and a model for sheltering obstacles. More on WAsP can be found at http://www.wasp.dk/.

Due to the scarcity of high-quality, ground-based solar resource measurements, large-area solar resource assessments in the U.S. have historically relied on the analysis of surface National Weather Service cloud cover observations. These observations are far more ubiquitous than solar measurements and allowed NREL to develop a 1961 to 1990 National Solar Radiation Database for 239 surface sites. However, more recently in the U.S., more and more reliance has been placed on Geostationary Satellite (GOES) visible channel data to obtain surface reflectance information that can be used to derive high-resolution (~10 km), site-time specific solar resource data (e.g., Perez et al., 2002). In fact, this approach has become commonplace in Europe, using Meteosat data. And the NASA Langley Research Center has recently completed a 20-year worldwide 100-km resolution surface solar energy dataset derived from International Satellite Cloud Climatology Project data, which is derived from data collected by all of the Earth's geostationary and polar orbiting satellites (http://eosweb.larc.nasa.gov/sse).

The use of satellite imagery for estimating surface solar resource characteristics over large areas has been studied for some years, and Renné et al. (1999) published a summary of approaches developed around the world. These satellite-derived assessments require good knowledge of the aerosol optical depth (AOD) over time and space, which can be obtained in part from MODIS and Advanced Very High Resolution Radiometer (AVHRR) data from polar orbiting environmental satellites. The assessments provide information both on Global Horizontal Irradiance, which is useful for estimating resources available to flat plate collectors such as photovoltaic panels or solar water heating systems, and Direct Normal Irradiance, which is needed for determining the resources available to solar concentrators that track the sun.

Besides NREL and NASA, other organizations perform similar types of high-resolution solar resource datasets. For example, the German Space Agency (DLR) has been applying similar methods to Meteosat data for developing solar resource maps and data for Europe and northern Africa. DLR was also involved in the SWERA project and applied their methodologies to several SWERA countries.

Geospatial Toolkit

Recently, NREL has begun to format the solar and wind resource information into GIS software-compatible formats and has incorporated this information, along with other geospatial data relevant to renewable energy development, into a Geospatial Toolkit (GsT). The GsT  is a standalone, downloadable, and executable software package that allows the user to overlay the wind and solar data with other geospatial datasets available for the region, such as transmission lines, transportation corridors, population (load) centers, locations of power plant facilities and substations, land use and land form data, terrain data, etc. Not only can the user overlay various datasets of their choosing, there are also simple queries built into the toolkit, such as the amount of "windy" land (e.g., class 3 and above) available within a distance of 10 km of all transmission lines (minus the specified exclusion areas, such as protected lands). The GsT developed at NREL makes use of the Environmental Science and Research Institute's (ESRI) Map Objects software, although other platforms, including online, Web-based platforms, could also be used.

In a sense, the GsT is a DSS, since it allows the user to manipulate resource information with other critical data relevant to the deployment of renewable energy technologies to assist decision makers in identifying and conducting preliminary assessments of possible sites for installing these systems and supporting renewable energy policy decisions. However, up to now, NREL has only prepared GsTs for a few locations: the countries of Sri Lanka, Afghanistan, and Pakistan; the Hebei Province in China; the state of Oaxaca in Mexico; and the state of Nevada in the U.S. By the time of publication of this chapter, additional toolkits may also be available. As with the resource data, all toolkits developed by NREL are available for download from NREL's Web site. Those toolkits developed under the SWERA project are also available from the SWERA Web site.

HOMER: NREL's Micropower Optimization Model

The primary DST that makes up the DSS being described here is HOMER, NREL's Micropower Optimization Model. HOMER is a computer model that simplifies the task of evaluating design options for both off-grid and grid-connected power systems for remote, standalone, and distributed generation applications. HOMER's optimization and sensitivity analysis algorithms allow the user to evaluate the economic and technical feasibility of a large number of technology options and to account for variation in technology costs and energy resource availability. HOMER can also address system component sizing and the adequacy of the available renewable energy resource. HOMER models both conventional and renewable energy technologies.

Power sources:

- Solar photovoltaic
- Wind turbine
- Run-of-river hydropower
- Generator: diesel, gasoline, biogas, alternative and custom fuels, and co-fired
- Electric utility grid
- Microturbine
- Fuel cell

Storage:

- Battery bank
- Hydrogen

Loads:

- Daily profiles with seasonal variation
- Deferrable (e.g., water pumping and refrigeration)
- Thermal (e.g., space heating and crop drying)
- Efficiency measures

In order to find the least cost combination of components that meet electrical and thermal loads, HOMER simulates thousands of system configurations, optimizes for lifecycle costs, and generates results of sensitivity analyses on most inputs. HOMER simulates the operation of each technology being examined by making energy balance calculations for each of the 8,760 hours in a year. For each hour, HOMER compares the electric and thermal load in the hour to the energy that the system can supply in that hour. For systems that include batteries or fuel-powered generators, HOMER also decides for each hour how to operate the generators and whether to charge or discharge the batteries. If the system meets the loads for the entire year, HOMER estimates the lifecycle cost of the system, accounting for the capital, replacement, operation and maintenance, fuel, and interest costs. The user can obtain screen views of hourly energy flows for each component as well as annual costs and performance summaries.

This and other information about HOMER are available on NREL's Web site: http://www.nrel.gov/homer/. The Web site also provides extensive

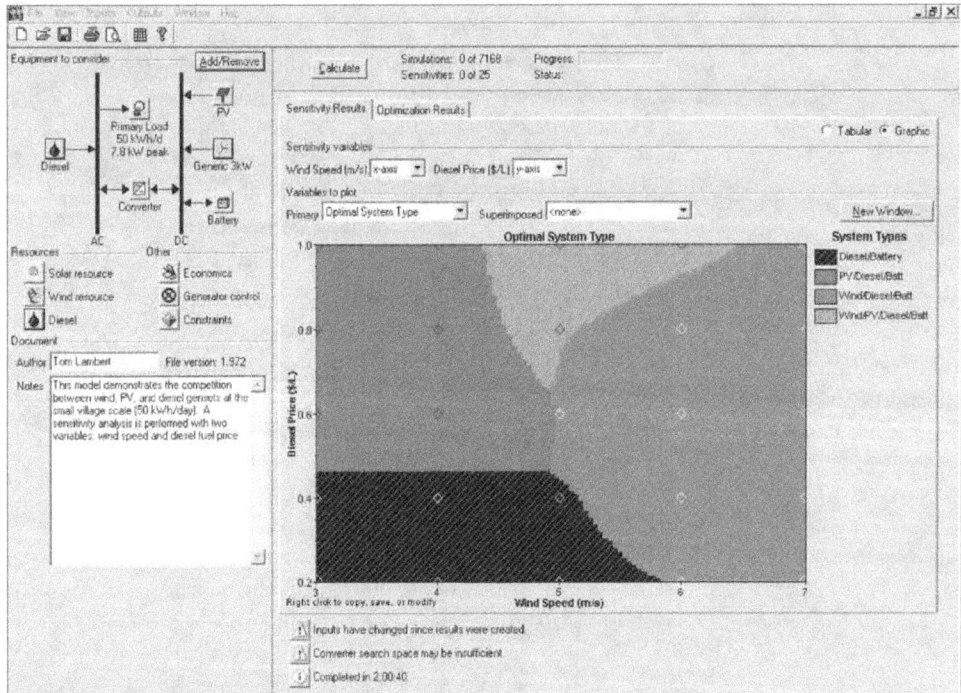

Figure 3-1 Example of HOMER output graphic. The column on the left provides a diagram showing the load characteristics and the types of equipment considered to meet the load. The optimal system design graphic shows the range within specified diesel fuel prices and wind energy resources for which various system types are most economical (e.g., a wind/diesel/battery system becomes the most optimal configuration to meet the load requirement for wind speeds greater than 5 m per second and fuel costs at 0.45 to 0.75$/l.

examples of how HOMER is used around the world to evaluate optimized hybrid renewable power systems to meet load requirements in remote villages. Figure 3-1 shows a typical example of an output graphic available from HOMER.

In order to accomplish these tasks, HOMER requires information on the hourly renewable energy resources available to the technologies being studied. However, typically hour-by-hour wind and solar data are not available for most sites. Thus, the user is requested to provide monthly or average information on solar and wind resources; HOMER then uses an internal weather generator to provide the best estimate of a simulated hour-by-hour dataset, taking into consideration diurnal variability if the user can provide an indication of what this should be. However, these approximations represent a source of uncertainty in the model. For those locations where a GsT is available, the GsT offers a mechanism for the user to easily ingest data from the toolkit into HOMER for the specific location of interest. However, since the toolkit contains only monthly solar and wind data, the limitations described above still apply. More information on the weather generator can be found in the HOMER Help files.

The HOMER developers have implemented various methods to facilitate access to reliable resource data that provide some of the input for simulations. For example, a direct link with the NASA surface meteorology and solar energy (SSE) data site enables the user to download monthly and annual solar data from any location on Earth. The 100-km resolution NASA data have become a benchmark of solar resource information due to the high quality of the modeling capability used to generate the data, the fact that the SSE is validated against numerous ground stations, and the fact that it is global in scope and now covers a 20-year period. However, the dataset is still limited by a somewhat coarse resolution and no validation in areas where ground data do not exist. The procedures used to generate the SSE also have problems where land-ocean interfaces occur, and in snow-covered areas.

Linking HOMER to higher-resolution regional solar datasets would likely improve these uncertainties somewhat, but in general, these datasets are also limited to monthly and seasonal values. However, since these methods rely on geostationary satellite data that provide frequent imagery of the Earth's surface, an opportunity exists to produce hourly time series data for up to several years at a 10-km resolution. This option will require significant data storage and retrieval capabilities on a server, but such a possibility now exists for future assessments.

Wind data available to HOMER is also generally limited to annual and, at best, monthly values. The standard HOMER interface allows the user to also designate a Weibull "k" value if this information is available. The Weibull k is a statistical means of defining the frequency distribution of the long-term hourly wind speeds at a location; this value can vary substantially depending on local terrain and microclimatic conditions. HOMER also has a provision for the user to designate the diurnal range of wind speeds and the timing when maximum and minimum winds occur. This information then provides improved simulation of the hour-by-hour wind values. The difficulty is that there may be applications where even these statistical values are not known to the user and are not available from the standard wind resource maps produced for a region, but this limitation may not be critical and requires further study to determine the impact on model output uncertainties.

2a. Access to the HOMER DSS

HOMER was originally developed and has always been maintained by the National Renewable Energy Laboratory (NREL). The model can be downloaded free of charge from NREL's Web site at http://www.nrel.gov/ homer/default.asp. The user is required to register, and registration must be updated every six months. The Web site also contains a variety of guides for getting started and using the software.

Resource information required as input to HOMER is generally freely available at the Web sites of the institutions developing the data. These institutions also generally maintain and continuously update the data. For example, renewable energy resource information can be found in several places on NREL's Web site, such as http://rredc.nrel.gov or www.nrel.gov/GIS. NASA solar energy data, which can be easily input to HOMER, is available at http://eosweb.larc.nasa.gov/sse. In fact, there is a specific feature built into HOMER that automatically accesses and inputs the SSE data for the specific location that the model is analyzing. Wind and solar resource data for the 13 SWERA countries can be found at http:// unep.swera.net. This Web site is currently undergoing expansion and upgrading by the USGS/EROS Data Center in Sioux Falls, South Dakota, and will eventually become a major clearing house for resource data from around the world in formats that can be readily ingested into tools such as HOMER.

2b. Definition of HOMER information requirements

The ideal input data format to HOMER is an hourly time series of wind and solar resource data covering a complete year (8,760 values). In addition, the wind data should be representative of the wind turbine hub height that is being analyzed within HOMER. Unfortunately, datasets such as these are seldom available at the specific locations for which HOMER is being applied. More typically, the HOMER user will have to identify input datasets from resource maps (even within the GsT, the resource data are based on what is incorporated into the map, which, in the case of wind, may represent only a single annual value). Because monthly and annual mean data are more typically available, HOMER has been designed to take monthly mean wind speeds (in meters/second) and monthly mean solar resource values (in kilowatts-hour/m2-day). In the case of wind, HOMER also allows for the specification of other statistical parameters related to wind speed distributions and diurnal characteristics. Furthermore, if the wind data available for input to HOMER do not represent the same height above the ground as the wind turbine's hub height being analyzed, HOMER has internal algorithms to adjust for this. The user must specify the height above the ground for which the data represent, and a power law conversion adjusts the wind speed value to the hub height of the specific wind turbine being analyzed. HOMER then utilizes an internal weather generator that takes the input information and creates an hour-by-hour data profile representing a one-year data file. Then, HOMER calculates turbine energy output by converting each hourly value to the energy production of the machine using the manufacturer's turbine power curve.

Besides the mean monthly wind speeds, the statistical parameters required by HOMER to generate the hourly datasets include the following:

- The altitude above sea level (to adjust for air density since turbine performance is typically rated at sea level);

- The Weibull k value, which typically ranges from 1.5 to 2.5 depending on terrain type;

- An auto-correlation factor, which is a measure of how strongly the wind speed in 1 hour depends (on average) on the wind speed in the previous hour (these values typically range from 0.85 to 0.90);

- A diurnal pattern strength, which is a measure of how strongly the wind speed depends on the time of day (values are typically 0.0 to 0.4); and

- The hour of the peak wind speed (over land areas, this is typically 1400 to 1600 local time).

In the U.S. as elsewhere, wind resource maps often depict the resource in terms of wind power density in units of watts-m-2 rather than in wind speeds. In this case, the wind power density must be converted back to a mean wind speed. The relationship between wind power density (P) and wind speed (v) is given as follows:

$$P = \tfrac{1}{2}\rho\sum_i v_i^3 \qquad (2)$$

where ρ is the density of the air and i is the individual hourly wind observation. Since the frequency distribution of wind speed over the period of a year or so follows a Weibull distribution shape, the wind power density can be converted back to a wind speed if the "k" factor in the Weibull distribution is known, as well as the height above sea level of the site (to determine the air density).

2c. Access to and use of the HOMER DSS among the federal, state, and local levels

Because of the easy access to HOMER and the related resource assessment data products, the HOMER DSS is freely available to all government and private entities in the U.S. and worldwide. Thousands of users from all economic sectors are using HOMER to evaluate renewable energy technology applications, particularly for off-grid use.

2d. Variation of the HOMER DSS by geographic region or characteristic

A key feature of HOMER is the evaluation of specific renewable energy technologies and related energy systems for different regions and for different applications. The HOMER model contains information on renewable energy technology characteristics; however, these characteristics, such as power curves for different wind turbine models, generator fuel curves, and other factors, are not affected by location. Because of the location-specific dependency of resource data, use of data that is not representative of the specific region of analysis will introduce additional uncertainties in the model results. Thus, the user should evaluate the accuracy and relevancy of any default information that is built into HOMER or any resource data chosen as input to HOMER before completing the final analyses.

3. Observations used by the HOMER DSS now and for potential use in the future

This section focuses on the Earth observations (of all types, from remote sensing and in situ) used or for potential use in the HOMER DSS.

3a. Kinds of observations being used

In the previous section, we provided a description of the renewable energy resource assessment related to solar and wind technologies that are required as input to HOMER when these technologies are being modeled. As noted in that section, developing this resource information requires the use of a variety of Earth observations. In this section, we list these observations for each resource category as well as other types of observations relevant to the HOMER DSS.

Wind Resources

The ideal observational platform for obtaining reliable wind resource data to be input into HOMER would be calibrated wind speed measurements from a meteorological tower installed at the location of interest. These measurements should be obtained at the hub height of the wind turbine being modeled, be of sufficient sampling frequency to provide hourly measurements, and be of sufficient quality and duration to result in at least one full year of continuous measurements. Although measurements of this quality are typically necessary at project sites where significant investments in large grid-connected wind turbines are anticipated, and where a decision has already been made to implement a large-scale project, it is extremely rare that this level of observation is available for most HOMER applications where the user is examining potential applications for proposed projects. Thus, some indirect means to establish wind characteristics at a proposed site, such as extrapolating wind resource measurements available from a nearby location or developing a wind resource map such as described in Section 2, is required. The major global datasets typically used by NREL for wind resource assessment are summarized in Table 3-1. More discussion on some of these datasets is provided below.

SURFACE STATION DATA

In the U.S., as well as in most other countries, the main source of routine surface wind observations would be observations from nearby national weather stations, such as those routinely maintained to support aircraft operations at airports. These data can be made available to the user from the NCDC in the form of the Integrated Surface Hourly (ISH) dataset. This database is composed of worldwide surface weather observations from about 20,000 stations that have been collected and stored from sources such as the Automated Weather Network, the Global Telecommunications System, the Automated Surface Observing System (ASOS), and data keyed from paper forms (see, http://gcmd.nasa.gov/records/GCMD_ C00532.html).

SATELLITE-DERIVED OCEAN WIND DATA

Ocean wind data can be obtained from the SeaWinds Scatterometer (see http://manati.orbit.nesdis.noaa.gov/ quikscat/) mounted aboard NASA's Quickscat satellite. Quickscat was launched on June 19, 1999 in a sun-synchronous polar orbit. A longer-term ocean winds dataset is available from the Special Sensor Microwave/ Imager (SSM/I) data products as part of NASA's Pathfinder Program. The SSM/I geophysical dataset consists of data derived from observations collected by SSM/I sensors carried onboard the series of Defense Meteorological Satellite Program polar orbiting satellites (see http://www.ssmi.com/ssmi/ssmi_description. html#ssmi). An example of how more recent QuickScat data were used in support of a wind resource assessment in Pakistan is provided in Figure 3-2 (see also http:// www.nrel.gov/applying_technologies/applying_ technologies_pakistan.html; click under "Monthly maps of satellite-derived wind speed estimates at 10 m above the surface for the Arabian Sea" at the Wind Resources section). Airborne or space-borne Synthetic Aperture Radar systems can also provide information on ocean

Table 3-1 Major Global Datasets Used by NREL for Wind Resource Assessment

Dataset	Type of Information	Source	Period of Record
Surface station data	Surface observations from more than 20,000 stations worldwide	National Oceanic and Atmospheric Administration NOAA/ National Climatic Data Center (NCDC)	Variable up to 2006
Upper air station data	Rawinsonde and pibal observations at 1,800 stations	NCAR	1973–2005
Satellite-derived ocean wind data	Wind speeds at 10 m above the ocean surface gridded to 0.250	NASA/Jet Propulsion Laboratory	1988–2006
Marine climatic atlas of the world	Gridded (1.00) statistics of historical ship wind observations	NOAA/NCDC	1854–1969
Reanalysis upper air data	Model-derived gridded (~200-km) upper air data	NCAR-NCEP	1958–2005
Global upper air climatic atlas	Model-derived gridded (2.50) upper air statistics	NOAA/NCDC	1980–1991
Digital geographic data	Political, hydrograph, etc.	ESRI	Not applicable (N/A)
Digital terrain data	Elevation at 1-km spatial resolution	USGS/EROS	N/A
Digital land cover data	Land use/cover and tree cover density at 0.5-km resolution	NASA/USGS	N/A

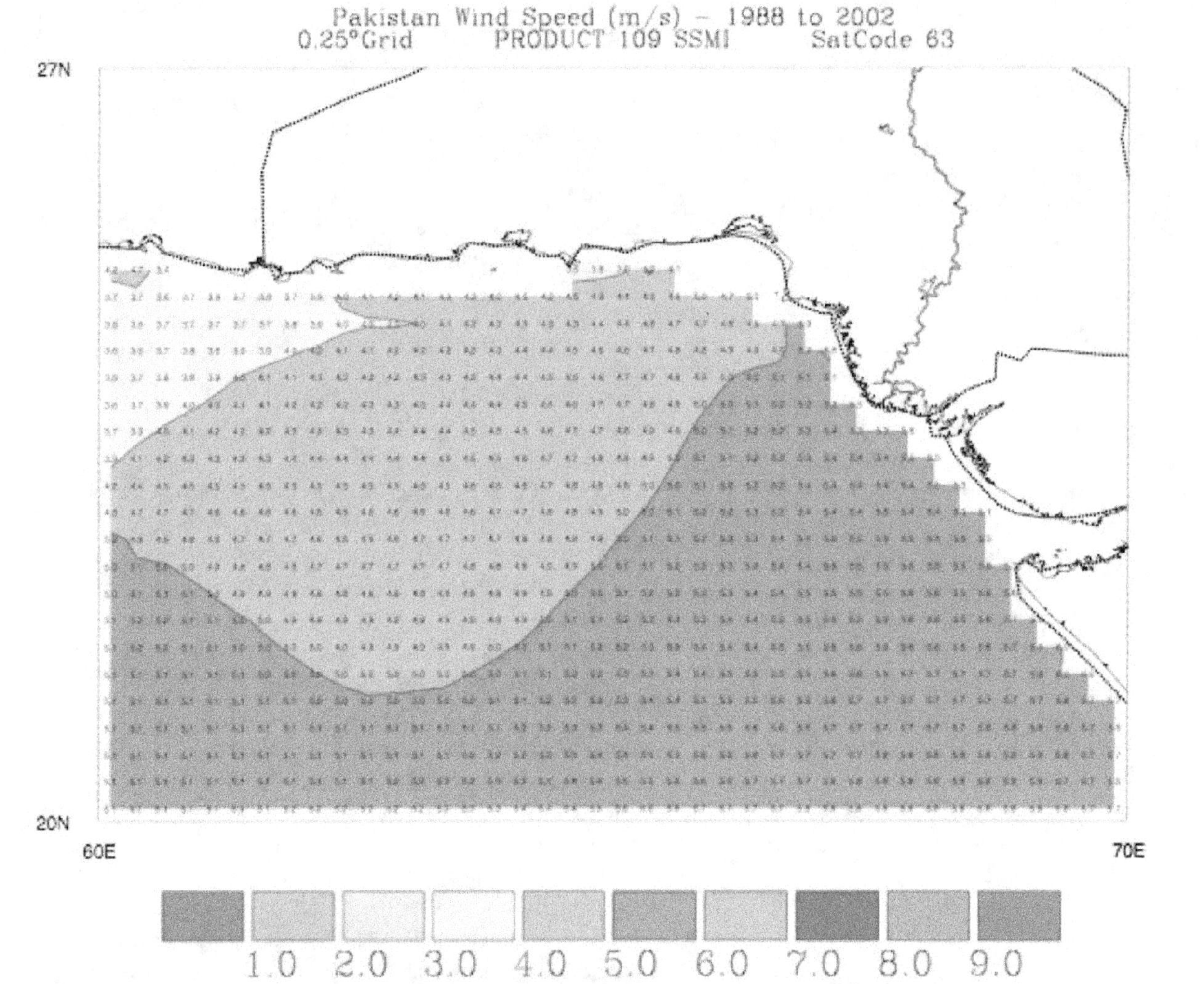

Pakistan Wind Speed (m/s) – 1988 to 2002
0.25°Grid   PRODUCT 109 SSMI   SatCode 63

Figure 3-2 Example of ocean wind resource assessment output for the offshore regions of Pakistan. These data were derived from the SeaWinds scatterometer aboard NASA's Quickscat satellite. The assessment provides estimated mean annual wind speeds at 10-m above the ocean surface averaged over the period from 1988 to 2002.

wind data, although these data are not commonly used for this purpose in the U.S. since scatterometer data products are more readily and freely available.

REANALYSIS OF UPPER AIR DATA

The U.S. reanalysis dataset was first made available in 1996 to provide gridded global upper air and vertical profiles of wind data derived from 1,800 radiosonde and pilot balloon observations stations (Kalnay et al. 1997). The reanalysis data were prepared by NCAR-NCEP and can be found at http://www.cdc.noaa.gov/cdc/reanalysis/. An early analysis of the dataset (Schwartz, George, and Elliott, 1999) showed that for wind resource assessments, the dataset was a promising tool for gaining a more complete understanding of vertical wind profiles around the world but that discrepancies with actual radiosonde observations still existed. Since that time, continuous improvements have been made to the NCAR-

NCEP dataset, and it is has become an ever-increasingly important data source for contributing to reliable wind resource mapping activities.

DIGITAL TERRAIN DATA

Digital Elevation Models (DEM) have been accessed from the USGS/EROS data center. These models consist of a raster grid of regularly spaced elevation values that have been derived primarily from the USGS topographic map series. The USGS no longer offers DEMs, and for the U.S., these can now be accessed from the National Elevation Dataset (http://ned.usgs.gov/). The SRTM offers much higher resolution terrain datasets, which are now beginning to be used in some wind mapping exercises. These are also being distributed by USGS/EROS under agreement with NASA (http://srtm.usgs.gov/).

DIGITAL LAND COVER DATA

Land cover data are used to estimate roughness length parameters required for the mesoscale meteorological models used in the wind mapping process. Data from the Global Land Cover Characterization dataset provide this information at a 1-km resolution (see http://edcsns17.cr.usgs.gov/glcc/background.html). The Moderate Imaging Spectroradiometer (MODIS) is used to obtain global percent tree cover values at a spatial resolution of 0.5 km (Hansen et al., 2003). Existing natural vegetation is also being mapped at a 200-m resolution as part of the USGS Regional Gap Analysis program. Gap analysis is a scientific method for identifying the degree to which native animal species and natural communities are represented in our present-day mix of conservation lands (Jennings and Scott, 1997).

Solar Resources

As with wind, the ideal solar resource dataset for incorporation into HOMER would be data derived from a quality, calibrated surface solar measurement system consisting of a pyranometer and a pyrheliometer that can provide a continuous stream of hourly data for at least one year. Such data are seldom available at the site for which HOMER is being applied. Although interpolation to nearby surface radiometer datasets can be accomplished with reasonable reliability, we usually resort to an estimation scheme to derive an in-situ dataset. The solar resource assessments that NREL and others undertake make use of several different observational datasets, such as ground-based cloud cover measurements, satellite-derived cloud cover measurements, or the use of the visible channel from satellite imagery data. The major global datasets used for solar resource assessments are summarized in Table 3-2. More discussion on some of these data products is described below.

WRDC

Since the early 1960s, the WRDC, located at the Main Geophysical Institute in St. Petersburg, Russia, has served as a clearinghouse for worldwide solar radiation measurements collected at national weather stations.

Table 3-2 Major global datasets used for solar resource assessments.

Dataset	Type of Information	Source	Period of Record
Surface station data	Surface cloud observations from more than 20,000 stations worldwide	NOAA/NCDC	Variable up to 2006
World Radiation Data Center (WRDC)	Surface radiation observations from over 1,000 stations worldwide	WRDC, St. Petersburg	1964–1993
Satellite imagers	Imagery from the visible channel of geostationary weather satellites (1-km resolution)	NASA/NOAA	1997– present
International Satellite Cloud Climatology Project	Used in the 10 global surface solar energy meteorological dataset	NASA/SSE	1983–2003
Aerosol RObotic NETwork (AERONET)	Observations of AOD from around the world	NASA/Goddard	Variable depending on station
Global Aerosol Climatology Project (GACP)	AODs (generally over oceans) at 10 x 10 from AVHRR data	NASA	1981–2005
MODIS, Multi-Angle Imaging Spectroradiometer (MISR), and Total Ozone Mapping Spectrometer (TOMS)	AOD	NASA	Variable since 1980s
Global Ozone Chemistry Aerosol Transport (GOCART)	AOD for turbid areas	NASA	March 30– May 3, 2001
Global Aerosol Dataset (GADS)	AOD derived from theoretical calculations and proxies		Compilation of measurements and models
Digital geographic data	Political, hydrography, etc.	ESRI	N/A
Digital terrain data	Elevation at 1-km spatial resolution	USGS/EROS	N/A
Digital land cover data	Land use/cover and tree cover density at 0.5-km resolution	NASA/USGS	N/A

The WRDC is under the auspices of the World Meteorological Organization. A Web-based dataset was developed by NREL in collaboration with the WRDC and can be accessed at http://wrdc-mgo.nrel.gov/. This data archive covers the period from 1964 to 1993. For more recent data, the user should go directly to the WRDC home page at http://wrdc.mgo.rssi.ru/.

AEROSOL OPTICAL DEPTHS

After clouds, atmospheric aerosols have the greatest impact on the distribution and characteristics of solar resources at the Earth's surface. However, routine in-situ observations of this parameter have only recently begun. Consequently, a variety of surface- and satellite-based observations are used to derive the best information possible of the temporal and spatial characteristics of the atmospheric AOD. The most prominent of the surface datasets is the AERONET (http://aeronet.gsfc.nasa.gov/), a network of automated multiwavelength sun photometers located around the world. This network also has links to other networks, where the data may be less reliable. AERONET data can be used to provide ground-truth data for different satellite sensors that have been launched on a variety of sun-synchronous orbiting platforms since the 1980s, such as TOMS, the Advanced Very High Resolution Radiometer (AVHRR), MODIS, and MISR, the latter two being mounted on NASA's Terra satellite. As noted by Gueymard (2003), determination of AOD from satellite observations is still subject to inaccuracies, particularly over land areas, due to a variety of problems such as insufficient cloud screening or interference with highly reflective surfaces. The GACP, established in 1998 as part of the NASA Radiation Sciences Program and the Global Energy and Water Experiment, has as its main objectives to analyze satellite radiance measurements and field observations in order to infer the global distribution of aerosols, their properties, and their seasonal and interannual variations and to perform advanced global and regional modeling studies of the aerosol formation, processing, and transport (http://gacp.giss.nasa.gov/).

Other sources of AOD data include the GOCART model (http://code916.gsfc.nasa.gov/People/Chin/gocartinfo. html), which is derived from a chemical transport model. An older dataset, GADS, which can be found at http://www.lrz-muenchen.de/~uh234an/www/radaer/gads. html, is a theoretical dataset providing aerosol properties averaged in space and time on a 50 x 50 grid (Koepke et al., 1997).

Other Renewable Energy Resources

Although the scope of this chapter focuses on wind and solar energy resources, it is evident that many of the Earth observation datasets listed above can apply to other renewable energy resources as well. For example, hydropower resources can be determined by analysis of high-resolution DEM data, along with knowledge of the rainfall amounts over specific watersheds and the land use characteristics of these watersheds. Biomass resource assessments can be enhanced through use of MODIS data as well as other weather-related data and through evaluation of MODIS and AVHRR data to determine the Normalized Vegetation Index.

3b. Limitations on the usefulness of observations

In the absence of direct solar and wind resource measurements at the location for which HOMER is being applied, the observations described in Section 3a, when used in the wind and solar resource mapping techniques described in Section 2, will together provide useful approximations of the data required as input to HOMER. However, the observations all have limitations in that they do not explicitly provide direct observation of the data value required for the mapping techniques but only approximations based on the use of algorithms to convert a signal into the parameter of interest. These limitations for some of these datasets can be summarized as follows:

Surface Station Data

These are generally not available at the specific locations at which HOMER would be applied, so interpolation is required. Furthermore, they generally do not have actual solar measurements but rather proxies for these measurements (i.e., cloud cover). The wind data are generally collected at 10 m above the ground or less, and the anemometer may not be in a well-exposed condition. When the station observations are derived from human observations, they represent samples of a few minutes duration every 1 or 3 hours; therefore, many of the observations are missing. For those stations that have switched from human observations to Automated Surface Observation Stations (ASOS), the means of observation have changed significantly from the human observations, representing a discontinuity in long-term records. Occasionally, the location of the station is changed without changing the station identification number, which can also cause a discontinuity in observations. Similarly, equipment changes can cause a discontinuity in observations.

Satellite-Derived Ocean Wind Data

These data are not based on direct observation of the wind speed at 10 m above the ocean surface but rather from an algorithm that infers wind speeds based on the wave height observations provided by the scatterometers or Synthetic Aperture Radar.

Satellite-Derived Cloud Cover and Solar Radiation Data

These datasets are derived from observations of the reflectance of the solar radiation from the Earth-atmosphere system. Although it could be argued that this method does provide a direct observation of clouds, the solar radiation values are determined from an algorithm that converts knowledge of the reflectance observation, the incoming solar radiation at the top of the atmosphere, and the transmissivity characteristics of the atmosphere to develop estimates of solar radiation.

Aerosol Optical Depth

Considerable research is underway to improve the algorithms used to convert multispectral imagery of the Earth's surface to AOD. The satellite-derived methods have additional shortcomings over land surfaces where irregular land-surface features make application of the algorithms complicated and uncertain.

3c. Reliability of the observations

For those observations that provide inputs to the solar and wind resource data, their reliability can vary from parameter to parameter. Generally all of the observations used to produce data values required for solar and wind assessments have undergone rigorous testing, evaluation, and validation. This research has been undertaken by a variety of institutions, including the institutions gathering the observations (e.g., NASA and NOAA) as well as the institutions incorporating the observations into resource mapping techniques (e.g., NREL). Many of the satellite-derived observations of critical parameters will be less reliable than in-situ observations; however, satellite-derived observations must still be used due to the scarcity of in-situ measurement stations.

3d. What kinds of observations could be useful in the near future

All of the observations currently available continue to be of critical value in the near future. For renewable energy resource mapping, improved observations of key weather parameters (wind speed and direction at various heights above the ground and over the open oceans at higher and higher spatial resolutions, improved ways of differentiating snow cover and bright reflecting surfaces from clouds, etc.) should be of value to the renewable energy community. New, more accurate methods of related parameters such as AOD would result in improvements in the resource data. All of these steps will lead to improvements in the quality of outputs from renewable energy DSTs such as HOMER.

4. Uncertainty

Application of the HOMER DSS involves a variety of input data types, all of which can have a level of uncertainty attached to them. HOMER addresses uncertainties by allowing the user to perform sensitivity analyses for any particular input variable or combination of variables. HOMER repeats its optimization process for each value of that variable and provides displays to allow the user to see how the results are affected. An input variable for which the user has specified multiple values is called a sensitivity variable, and users can define as many of these variables as they wish. In HOMER, a "one-dimensional" sensitivity analysis is done if there is a single sensitivity variable, such as the mean monthly wind speed. If there are two or more sensitivity variables, the sensitivity analysis is "two" or "multidimensional." HOMER has powerful graphical capabilities to allow the user to examine the results of sensitivity analyses of two or more dimensions. This is important for the decision maker, who must factor in the uncertainties of input variables in order to make a final judgment on the outputs of the model.

The amount of uncertainty associated with resource data is largely dependent on how the data are obtained and on the nature of the analysis being undertaken. For some types of analyses, very rough estimates of the wind resource would be sufficient; for others, detailed hourly average data based on surface measurements would be necessary. Quality in-situ measurements of wind and solar data in formats suitable for renewable energy applications over a sufficient period of time (one year or more) can have uncertainties of less than ±3% of the true value. However, when estimation methods are required, such as the use of Earth observations and modeling and empirical techniques, uncertainties can be as much as ±10% or more. These uncertainties are highest for shorter-term datasets and are lower when annual average values are being used since, throughout the year, errors in the estimation methods have a tendency to compensate among the individual values.

Based on wind turbine and solar technology operating characteristics, it is possible that the error in estimating a renewable energy system performance over a year is roughly linear to the error in the input resource data. For example, for wind energy systems, even though the power of the wind available to a wind turbine is a function of the cube of the wind speed, it turns out that the turbine operating characteristics are such that wind turbines typically do not produce any power at all until a certain threshold speed is reached, at which point the power output increases approximately linearly with wind speed until the winds are so high that the turbine must shut down. This results in an annual turbine power output that is roughly linear to the mean annual wind speed for certain mean wind speed ranges. This would mean that, in some cases, an uncertainty in the annual wind or solar resource of ±10% results in an uncertainty of expected renewable energy technology output of approximately ±10%.

5. Global change information and the HOMER DSS

This section expands the discussion of the HOMER DSS to include the relationship of HOMER and its input data requirements with global change information.

5a. Reliance of HOMER DSS global change information

As shown in the previous section, a number of observations that provide information on global change are also used in either direct or indirect ways as input to HOMER. These observations relate primarily to the renewable energy resource information that is required for HOMER applications. Renewable energy system performance is highly dependent on the local energy resources available to the technologies. The extent and characteristics of these resources are driven by weather and local climate conditions, which happen to be the primary areas that Earth observational systems monitoring climate change are addressing. Thus, as users seek access to observations to support renewable energy resource assessments, they will invariably be seeking certain global change observational data.

Specifically, users will be seeking global change data related to atmospheric properties that support the assessment of solar and wind energy resources, such as wind and solar data and atmospheric parameters important for estimating these data. For example, major datasets used in solar and wind energy assessments include long-term reanalysis data, climatological surface weather observations, and a variety of satellite observations from both active and passive onboard remote sensors.

Key factors in affecting the choice of these observational data are their relevance to conducting reliable solar and wind energy resource assessment, their ease of access, and low or no cost to the user. The extensive list of observational data being used in the assessment of renewable energy resources represents strong leveraging of major taxpayer-supported observational programs that are geared primarily for global change assessment.

There is also an important consideration regarding the potential influence of long-term climate change

on the renewable energy resources that are used as input into HOMER. Through the Intergovernmental Panel on Climate Change, there has been a significant improvement in the reliability and spatial resolution of General Circulation Models (GCM) used to estimate the impacts of greenhouse gas emissions on climate change. As weather patterns change under changing climate conditions, wind and solar energy resources at a specific location can also change over time. The GCM results indicate that these renewable energy resources can be measurably different in 50 to 100 years from now than they are today in specific locations and regions. These changes may have a noticeable impact on the results of HOMER simulations in the future; however, significant uncertainties exist in GCM results. Until these uncertainties are reduced sufficiently, implementation of GCM results will produce unreliable HOMER simulations.

5b. How the HOMER DSS can support climate-related management decision-making among U.S. government agencies

Although HOMER was not intentionally designed to be a climate-related, management decision-making tool, the HOMER DSS has attributes that can support these decisions. For example, as we explore mechanisms for mitigating the growth of carbon emissions in the atmosphere, the HOMER DSS can be deployed to evaluate how renewable energy systems can be used cost-effectively to displace energy systems dependent on fossil fuels. Clearly, the science results, global change data, and information products coming out of our reanalysis and satellite-borne programs are of critical importance to HOMER for supporting this decision-making process. Given that the pertinent observational datasets have been developed primarily by federal agencies, these datasets tend to be freely available or available at a relatively small cost given the costs involved in making the observations in the first place. However, as we have noted in previous sections, the use of global change observations as input to the resource assessment data required by HOMER is not the optimal choice of data; ideally, in-situ (site-specific) measurements of wind and solar data relevant to the technologies being analyzed would be the most useful and accurate data to have for HOMER, if they were available.

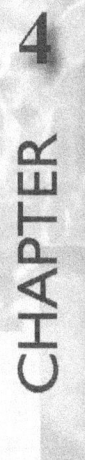

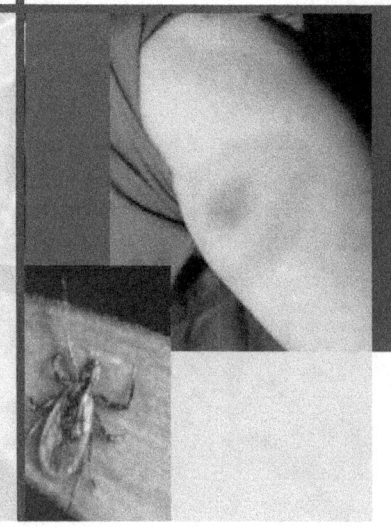

Decision Support for Public Health

Lead Author: Gregory E. Glass

1. Introduction

Public health is an approach to protect and improve the health of community members by preventive medicine, health education, control of communicable diseases, application of sanitary measures, and monitoring and control of environmental hazards (Lilienfeld and Lilienfeld 1980). This overall task is achieved by conducting basic and applied biomedical research to identify and ameliorate adverse health impacts, assessing and monitoring populations at risk to identify health problems, and establishing priorities to formulate policies to solve identified problems and to ensure populations have access to appropriate care, including health promotion, disease prevention, and evaluation of care. During the past century, the notable public health achievements as identified by the United States (U.S.) Centers for Disease Control and Prevention (CDC) include vaccinations and treatments against infectious diseases, injury prevention strategies, reduced occupational exposures to toxins, improved food and water safety, decreases in childhood and maternal mortality, and safer water sources. Thus, many of the key issues related to public health are incorporated in previous chapters in this report, though they may not be characterized as public health. Regardless, public health may represent a key dimension in problem solving under climate change situations. Many of the anticipated public health consequences of climate change are due to the influences of temperature and precipitation patterns. These factors, as well as attendant land cover changes, are expected to affect human communities. For example, changes in the availability of food resources and the quality of drinking water are anticipated to directly affect nutritional status and the spread of communicable infectious agents. In extreme situations, these conditions create "environmental refugees"—individuals displaced by serious environmental changes such as rising sea levels, desertification, dried up aquifers, weather-induced flooding, and other climatic changes (Huntingford et al., 2007).

Because public health is an important outcome component of decision-support tools (DST) involving air quality, water management, energy management, and agricultural efficiency issues, this chapter will focus on a unique public health aspect of DST/Decision-Support System (DSS) by examining infectious disease systems. Infectious diseases remain a significant burden to populations both globally as well as within the U.S. Some of these, such as syphilis and measles, involve a relatively simple dynamic of the human host population and the parasite—be it a virus, bacterium, or other micro-organism. These diseases, therefore, tend to be influenced by social behavior and the ability to provide resources and health education to significantly alter human behavior. However, other disease systems include additional species for their successful transmission— either wildlife species that maintain the micro-organism (zoonoses) or insect or arthropod vectors that serve to transmit the parasites either among people or from wildlife to people (vector-borne diseases).

Some of the most significant diseases globally are vector-borne or zoonotic diseases. Examples include malaria and dengue. In addition, many newly recognized (i.e., emerging) diseases either are zoonoses, such as Severe Acute Respiratory Syndrome (SARS), or appear to have been derived from zoonoses that became established in human populations (e.g., Human Immunodeficiency Virus [HIV]). Changes in rates of contact between component populations of these disease systems alter the rates of infectious disease (Glass, 2007). Many of these changes come about through activities involving the movement of human populations into areas where these pathogen systems normally occur or when they can occur because people introduce materials with infectious agents into areas where they were not known previously (Gubler et al., 2001). The introduction of West Nile virus from its endemic area in Africa, the Middle East, and Eastern Europe into North America and its subsequent spread across the continent is a recent example. The impacts of

the virus on wildlife, human, and agricultural production are an excellent example of the economic consequence of such emergent disease systems.

More recently, attention has focused on the potential impact that climate change could have on infectious disease systems, especially those with vector or zoonotic components (e.g., Gubler et al., 2001). Alterations in climate could impact the abundances or interactions of vector and reservoir populations or the way in which human populations interact with them (Gubler, 2004). In addition, there is speculation that climate change will alter the locations where disease systems are established, shifting the human population that is at risk from these infectious diseases (e.g., Brownstein et al., 2005a; Fox, 2007).

Unlike many of the other applications in this report where Earth observations and modeling are of growing importance, the use of Earth observations by the public health community has been sporadic and incomplete. Although early demonstrations showed their utility for identifying locations and times that vector-borne diseases were likely to occur (e.g., Linthicum et al., 1987; Beck et al.,1997), growth of their application has been comparatively slow. Details of the barriers to implementation include the need to "scavenge" data from Earth observation platforms as none of these are designed for monitoring disease risk. This is not an insurmountable problem and, in fact, only a few applications for Earth observations have dedicated sensors. However, disease monitoring requires a long history of recorded data to provide information concerning the changes in population distribution and the environmental conditions associated with outbreaks of disease. Detailed spectral and spatial data need to be of sufficient resolution, and the frequency of observations must be high enough to enable identification of changing conditions (Glass, 2007). As a consequence, many DSTs undergoing development have substantial integration of Earth observations but lack an end-to-end public health outcome, particularly when focusing on infectious diseases. Therefore, the DSS to Prevent Lyme Disease (DSSPL), supported by the CDC and Yale University, was selected to demonstrate the potential utility of these systems within the context of climate change science. Lyme disease is a vector-borne, zoonotic bacterial disease. In the U.S., it is caused by the spirochete, Borrelia burgdorferi, and it is the most common vector-borne disease with tens of thousands of reported cases annually (Piesman and Gern 2004). Most human cases occur in the eastern and upper Mid-West portions of the U.S., although there is a secondary focus along the West Coast of the country. In the primary focus, the black-legged tick (or deer tick) of the genus Ixodes is most often found infected with B. burgdorferi.

2. Description of DSSPL

The diverse ways in which Lyme disease presents itself in different people has made it a public health challenge to ensure that proper priorities are established, to formulate policies to solve the problem, and to ensure that populations have access to appropriate care. The CDC uses DSSPL to address questions related to the likely distribution of Lyme disease east of the 100th meridian, where most cases occur (Brownstein et al., 2003). This is done by identifying the likely geographic distribution of the primary tick vector (the black-legged tick) in this region. DSSPL uses field reports of the known distribution of collected tick vectors, as well as sites with repeated sampling without ticks as the outcome space. DSSPL uses satellite data and derived products, such as land cover characteristics, census boundary files, and meteorological data files, to identify the best statistical predictor of the presence of black-legged ticks within the region. Land cover is derived from multi-date land remote-sensing satellite (Landsat) telemetry imagery and 10-meter (m) panchromatic imagery.

DSSPL combines the satellite and climate data with the field survey data of Ixodes ticks from at locally sampled sites throughout the region (Brownstein et al., 2003) or from rates of reported cases of Lyme disease (Brownstein et al., 2005b) in spatially explicit statistical models to generate assessment products of the distribution of the tick vector or human disease risk, respectively. These models are validated by field surveys in additional areas and the sensitivity and specificity of the results determined (Figure 4-1). Thus, the DSSPL is primarily a DST for prioritizing the likely geographic extent of the primary vector of Lyme disease in this region (Figures 4-1 and 4-2). It currently stops short of characterizing the risk of disease in the human population but is intended to delimit the area within which Lyme disease (and other diseases caused by additional pathogens carried by the ticks) might occur (Figure 4-2). Researchers at Yale University are responsible for developing and validating appropriate analytical methods to develop interpretations that can deal with many of the challenges of spatially structured data, as well as the acquisition of Earth science data that are used for model DSSPL predictions. The distinction between the presence/abundance of the tick vector and actual human risk relies on the effects of human population abundance and behavioral heterogeneity (e.g., work or recreational activity) that can alter the contact rate between the tick vector and susceptible humans. However, such detailed human studies (especially behavioral heterogeneity) are typically not available (Malouin et al., 2003). In Brownstein et al. (2005b), they found that although the entomological risk (the abundance of infected ticks) increased with

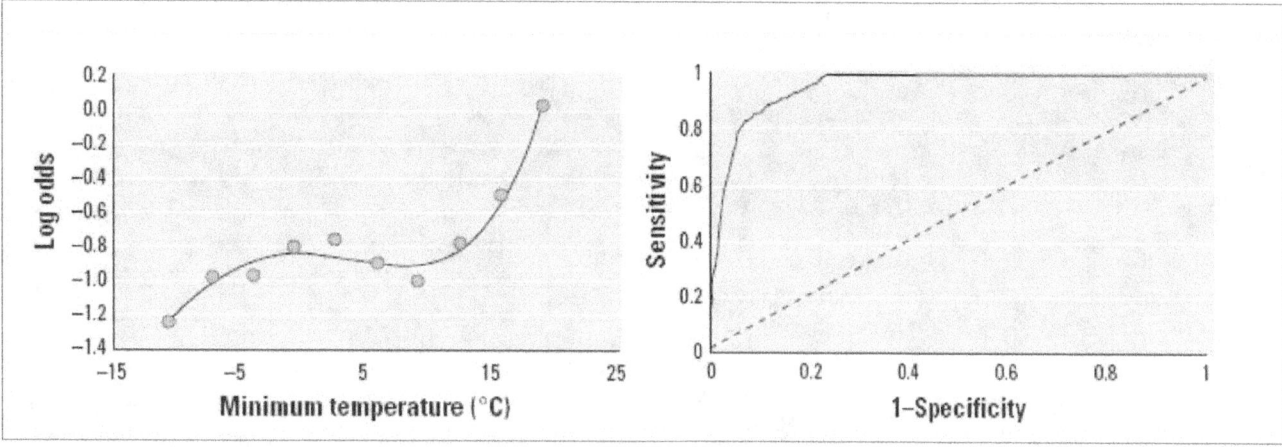

Figure 4-1 Relationship between the occurrence of black-legged tick presence at a site and minimum temperature (top) and evaluation of model (bottom) (Brownstein et al., 2003, Environmental Health Prospectus).
Left Panel: Log odds plot for relationship between I. Scapularis population maintenance and minimum temperature (T). Minimum temperature showed a strong positive association with odds of an established I. Scapularis population. According to the goodness of fit testing, the relationship was fit best by a fourth order polynomial regression (R2 = 0.97); log odds = 0.00000674 + 0.000273-0.0027T2 + 0.0002T − 0.8412.
Right Panel: Receiver operator characteristic plot describing the accuracy of the auto-logistic model. This method graphs sensitivity versus 1-specificity over all possible cutoff probabilities. The area under the curve is a measure of overall fit, where 0.5 (a 1:1 line) indicates a chance performance (dashed line). The plot for the auto-logistic model significantly outperformed the chance model with an accuracy of 0.95 (p<0.00005).

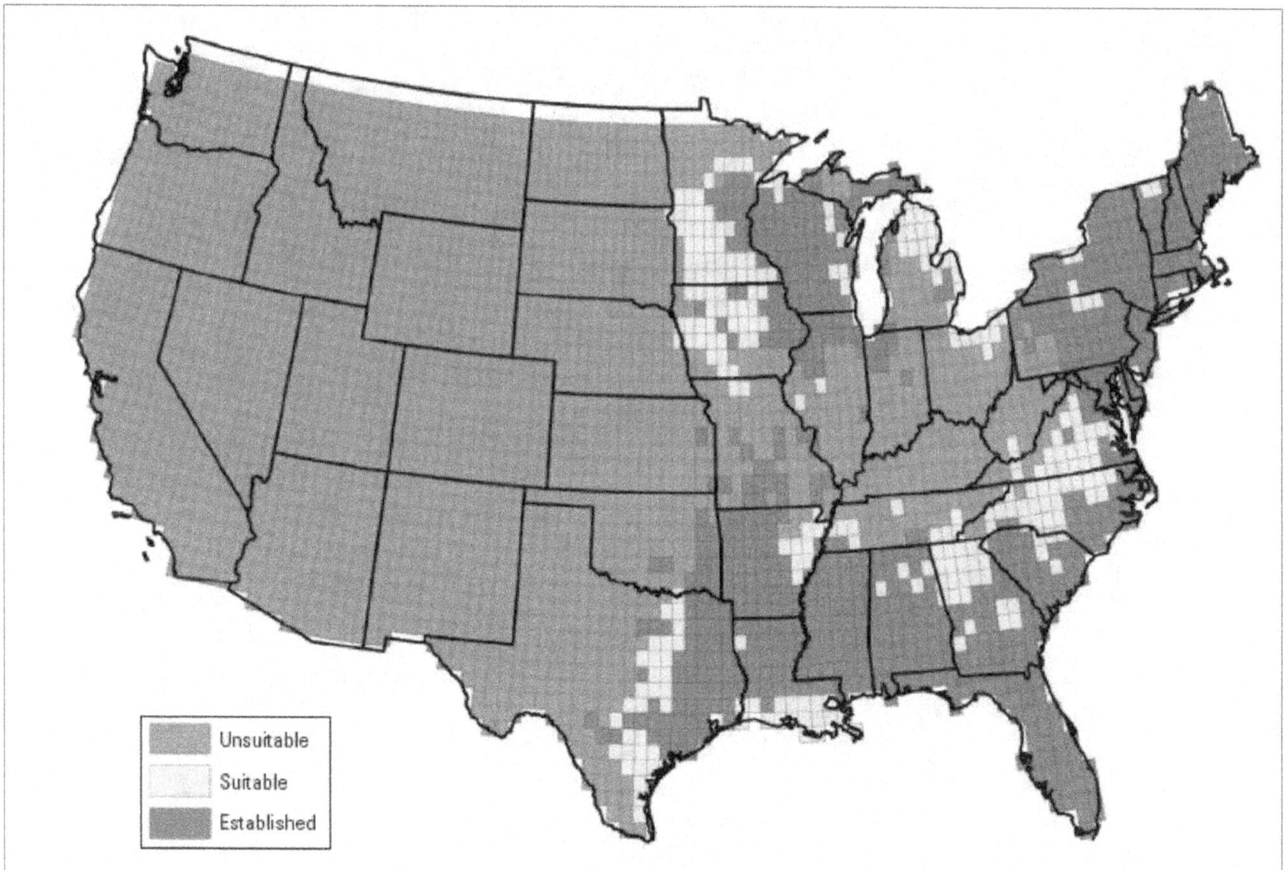

Figure 4-2 Forecast geographic distribution of the black-legged tick vector east of the 100th meridian in the U.S. for DSSPL (Brownstein et al., 2003, Environmental Health Prospectus 2a. New distribution map for I. Scapularis in the U.S.). To determine whether a given cell can support I. Scapularis populations, a probability cutoff point for habitat suitability from the auto-logistic model was assessed by sensitivity analysis. A threshold of 21% probability of establishment was selected, giving a sensitivity of 97% and a specificity of 86%. This cutoff was used to reclassify the reported distribution map (Dennis et al., 1998). The auto-logistic model defined 81% of the reported locations (n=427) as established and 14% of the absent areas (n=2,327) as suitable. All other reported and absent areas were considered unsuitable. All areas previously defined as established maintained the same classification.

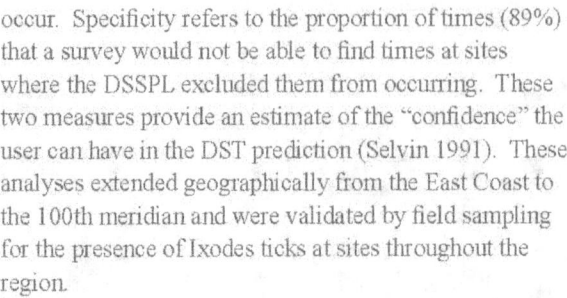

landscape fragmentation, the human incidence of Lyme disease decreased, thus indicating that there is a complex relationship between the landscape, the population of ticks, and the human response resulting in the health outcome.

3. Potential Future Use and Limits

Future use of DSSPL depends to a great extent on public health policy decisions exterior to the DST. The perspective of the role that Lyme disease prevention rather than treatment of diseased individuals will play is a key aspect of the importance that DSSPL will experience. For example, studies have shown that even in Lyme disease endemic regions, risk communication often fails to reduce the likelihood of infection (Malouin et al., 2003). In principle, policy makers may decide that it is more cost effective to provide improved treatment modalities rather than investing in educational programs that fail to reduce the disease burden. Alternatively, the development of vaccines is time consuming, costly, and may have additional risks of unacceptable side effects that affect the likelihood that this would be a policy choice. Thus, depending on policy decisions and the effects of alternative interventions, the DSSPL might be used to forecast risk areas for educational interventions, inform health care providers in making diagnoses, or plan mass vaccination campaigns.

Currently, the removal of the licensed Lyme disease vaccine from the general public has eliminated this as a strategy to reduce the disease burden. The apparent lack of impact of targeted education also makes this a less likely strategy. Thus, the extent to which treatment modalities rather than prevention of infection will drive the public health response in the near future will play a major role in the use of DSSPL. However, even if the decision is made to focus on treatment of potentially infected individuals, DSSPL may still be useful by identifying regions where disease risk may be low, helping health care workers to focus clinical diagnoses on alternate causes.

Presuming that the DST continues to be used, the need for alternative/improved Earth science data to clarify environmental data for DSSPL, such as land cover, temperature, and moisture regimes is currently uncertain. The present system reports a sensitivity of 88% and a specificity of 89%—generally considered a highly satisfactory result. Sensitivity and specificity are considered the two primary measures of a method's validity in public health analyses. Sensitivity in the DSSPL model refers to the expected proportion of times (88%) that ticks would be found when field surveys were conducted at sites that the DSSPL predicted they should

occur. Specificity refers to the proportion of times (89%) that a survey would not be able to find times at sites where the DSSPL excluded them from occurring. These two measures provide an estimate of the "confidence" the user can have in the DST prediction (Selvin 1991). These analyses extended geographically from the East Coast to the 100th meridian and were validated by field sampling for the presence of Ixodes ticks at sites throughout the region.

Typically, patterns of weather regimes appear to have a greater impact on distribution than more detailed information on land cover patterns. However, some studies indicate that fragmentation of forest cover and landscape distribution at fairly fine spatial resolution can substantially alter patterns of human disease risk (Brownstein et al., 2005b). These results also suggest that human incidence of disease may, in some areas of high transmission, be decoupled from the model constructed for vector abundance, reemphasizing the distinction between a key component (the vector) and actual human risk. When coupled with the stated accuracy of the DSSPL in identifying vector distribution, this would suggest that future efforts will probably require an additional model structure that includes sociological/behavioral factors of the human population that put it at varying degrees of risk. An additional limit of the DSSPL is that it does not explicitly incorporate human health outcomes in its analyses. In part, this reflects a public health infrastructure issue that limits detailed information on the distribution of human disease to (typically) local and state health agencies. For example, confidentiality of health records, including detailed locational data such as home addresses, are often shielded in the absence of explicit permission. This makes establishing the relationship between monitored environmental conditions and human health outcomes difficult. One solution is to aggregate data to some jurisdictional level. However, this produces the well known "ecological fallacy" in establishing relationships between environmental factors and health outcomes (Selvin 1991). With appropriate planning or the movement of the technology into local public health agencies, these challenges could be overcome. Some localized data (e.g., Brownstein et al., 2005b) of human health outcomes have been used to evaluate the utility of DSSPL and indicate that there is good potential for the DSS to provide important information on local risk factors.

4. Uncertainty

Uncertainty in decision making from DSSPL is based on the results of statistical analyses in which standard statistical models with spatially explicit components, such as autologistic intercepts of logistic models, are used to account for spatial autocorrelation in outcomes. The statistical analyses are well-supported theoretically. Typical calibration approaches involve model construction followed by in-field validation. Accuracy of classification is then assessed in a sensitivity-specificity paradigm.

However, little attention is paid in the current model to assessing uncertainty in the environmental data obtained from remotely sensed (or even in-situ) monitors of the environment. For example, most of the derivative data, such as land cover, may change with population growth and development. In addition, the use of average environmental conditions provide an approximate characterization of local, edaphic conditions that may affect the abundance of the tick vectors.

Whether these are the primary sources of "error" in the sensitivity and specificity results (although these are considered excellent results) of the DSSPL is not addressed and is an area the public health applications need to consider in future applications. Alternatively, there are biological reasons for the errors in the model, including the interaction of climatic factors and tick activity that may be responsible for sites predicted to have ticks that were not found to have them. To resolve some of the biological/environmental issues, validation is ongoing.

There also are a number of public health issues that affect the certainty of the DSSPL (and any DST) that are extrinsic to the system or tool. Accuracy in clinical diagnoses (both false positives and negatives), as well as reporting accuracy can affect the evaluation of the tool's utility. Currently, this is an issue of serious contention and forms part of the rationale for focusing on accurately identifying the distribution of the primary tick vector as an integral step in delimiting the distribution of the disease and evaluating the needs for the community.

5. Global Change Information and DSSPL

The relationship between climate and public health outcomes is complex. It is affected both by the direction and strength of the relationship between climatic variability and the component populations that make up a disease system, as well as the human response to changes in disease risk (Gubler 2004).

The DSSPL is one of the few public health DSTs that

has explicitly evaluated the potential impact of climate change scenarios on this infectious disease system. Assuming that evolutionary responses of the black-legged tick, B. burgdoferi, and the reservoir zoonotic species remains little changed under rapid climate change, Brownstein et al., (2005a) evaluated anticipated changes in the distribution and extent of disease risk.

This analysis used the basic climate-land cover suitability model developed for DSSPL and selected the Canadian Global Coupled Model (CGCM1) under two historically forced integrations. The first with a 1% per year increase in greenhouse gas emissions and the second with greenhouse gas and sulfate aerosol changes resulted in a 4.9 and 3.8□ Celsius increase in global mean temperature by the year 2080. Near- (2020), mid- (2050), and far-point (2080) outcomes were evaluated (Figure 4-3). The choice of CGCM1 was based on the Intergovernmental Panel on Climate Change criteria for vintage, resolution, and validity (Brownstein et al., 2005a).

Extrapolation of the analyses suggest that the tick vector will experience a significant range expansion into Canada but will also experience a likely loss of habitat range in the current southern portion of its range (Figure 4-3). This loss of range is thought to be due to impact of increased temperatures causing decreased survival in ticks when they are off their feeding hosts. It also is anticipated that its range will shift in the central region of North America, where it is currently absent. When coupled with the anticipated, continued human movement to more southern portions of the country, the numbers of human cases are expected to show an overall small decrease.

These long-range forecasts disguise a more dynamic process with ranges initially decreasing during near- and mid-term timeframes. This range reduction is later reversed in the long-term producing the overall pattern described by the authors. The impact in range distribution also produces an overall decrease in human disease risk as suitable areas move from areas of primary human concentration to areas that are anticipated to be less well populated.

Thus, DSS similar to those developed for Lyme disease have the potential for providing both near- and far-term forecasts of potential infectious disease risk that are important for public health planning. In addition, detailed studies (e.g., Brownstein et al., 2005b) provide public health agencies with important information on drivers of human risk that have been difficult to obtain by other means. As a consequence, DSS using remotely sensed data sources, either in part or whole, have the potential to significantly improve the health of communities.

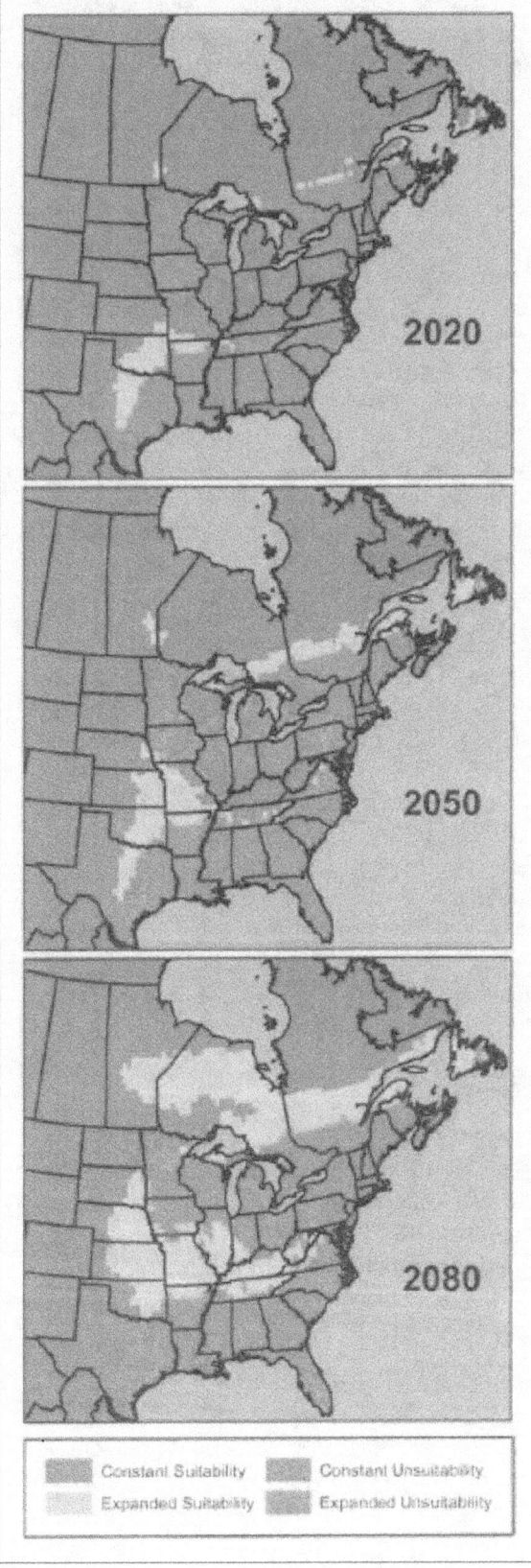

The primary challenges for the Earth science community involve understanding the needs of the public health community for the appropriate data at the appropriate spatial, temporal, and spectral scales. This will involve understanding a historically entrenched set of methodologies for interpreting health data and establishing causal relationships between inputs (environmental data) and outputs (health outcomes). In addition, there is the challenge of performing these tasks in the presence of limited resources for a community that has little cultural understanding of both the strengths and limitations of the data derived from these sources.

Figure 4-3 Forecast change in black-legged tick distribution in eastern and central North America under climate change scenarios using DSSPL (Brownstein et al., 2005a, EcoHealth)

CHAPTER 5

Decision Support for Water Resources Management

Lead Author: Holly C. Hartmann

1. Introduction

Water resource managers have long been incorporating information related to climate in their decisions. The tremendous, regionally ubiquitous investments in infrastructure to reduce flooding (e.g., levees and reservoirs) or assure reliable water supplies (e.g., reservoirs, groundwater development, irrigation systems, water allocation, and transfer agreements) reflect societal goals to mitigate the impacts of climate variability at multiple time and space scales. As the financial, political, social, and environmental costs of infrastructure options have become less tractable, water management institutions have undergone comprehensive reform, shifting their focus to optimizing operations of existing projects and managing increasingly diverse and often conflicting demands on the services provided by water resources (Bureau of Reclamation, 1992; Beard, 1993; Congressional Budget Office, 1997; Stakhiv, 2003; National Research Council [NRC], 2004). Governments have also made substantial investments to improve climate information and understanding over the past decades through satellites, in-situ measuring networks, supercomputers, and research programs. National and international programs have explicitly identified as an important objective ensuring that improved data products, conceptual models, and predictions are useful to the water resources management community (Endreny et al., 2003; Lawford et al., 2005). Although exact accounting is difficult, potential values associated with appropriate use of accurate hydrometeorologic predictions generally range from the millions to the billions of dollars (e.g., National Hydrologic Warning Council, 2002). There are also non-monetary values associated with more efficient, equitable, and environmentally sustainable decisions related to water resources.

Droughts, floods, and increasing demands on available water supplies continue to create concern and even crises for water resources management. Many communities have faced multiple hydrologic events that were earlier thought to have low probabilities of occurrence (e.g., NRC, 1995), and long-term shifts in streamflows have been observed (Lettenmaier et al., 1994; Lins and Slack, 1999; Douglas et al., 2000), leading to questions about the relative impacts of shifts in river hydraulics, land use, and climate conditions.

Until the last two decades, climate was viewed largely as a collection of random processes, and this paradigm informed much of the water resource management practices developed over the past 50 years that persist today. However, climate is now recognized as a chaotic process, shifting among distinct regimes with statistically significant differences in average conditions and variability (Hansen et al., 1997). As instrumental records have grown longer and extremely long time-series of paleoclimatological indicators have been developed (Ekwurzal, 2005), they increasingly belie one of the fundamental assumptions behind most extant water resources management—stationarity. Stationary time series have time-invariant statistical characteristics (e.g., mean or variance), meaning that different parts of the historical record can be considered equally likely. Within the limits posed by sampling, statistics computed from stationary time series can be used to define a probability distribution that will also then faithfully represent expectations for the future (Salas, 1993).

Further, prospects for climate change due to global warming have moved from the realm of speculation to general acceptance (Intergovernmental Panel on Climate Change [IPCC] 1990, 1995a, 2001a, 2007). The potential impacts of climate on water resources and their implications for management have been central topics of concern in climate change assessments (e.g., Environmental Protection Agency [EPA], 1989; IPCC, 1995b, 2001b; National Assessment Synthesis Team, 2000; Gleick and Adams, 2000; Barnett et al., 2004). These studies are becoming increasingly confident in

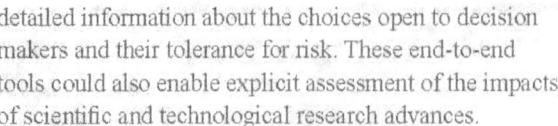

their conclusions that the future portends statistically significant changes in hydroclimatic averages and variability.

There has been persistent and broad disappointment in the extent to which improvements in hydroclimatic science from large-scale research programs have affected resource management practices in general (Pielke, 1995, 2001; NRC, 1998a, 1999a) and water resource management in particular (NRC, 1998b, 1999b,c). For example, seasonal climate outlooks have been slow to be entered into the water management decision processes, even though they have improved greatly over the past 20 years (Hartmann et al., 2002a, 2003). Water mangers have been even more resistant to incorporating notions of hydrologic non-stationarity in general and climate change in particular in decision processes. Until recently, hydrologic analysis techniques have been seen as generally sufficient (e.g., Matalas, 1997; Lins and Stakhiv, 1998), especially in the context of slow policy and institutional evolution (Stakhiv, 2003). However, an inescapable message for the water resource management community is the inappropriateness of the stationarity assumption in the face of climate change.

Several ongoing efforts are leading the way forward to establish more effective ways of incorporating climate understanding and earth observations into water resources management (Pulwarty, 2002; Office of Global Programs, 2004; NASA, 2005). While diverse in their details, these efforts seek to link hydroclimatological variability, analytical and predictive technologies, and water management decisions within an end-to-end context extending from observational data through large-scale analyses and predictions, uncertainty evaluation, impacts assessment, applications, and evaluations of applications (e.g., Young, 1995; Miles et al., 2000). Some end-to-end efforts focus on cultivating information and management networks; designing processes for recurrent interaction among research, operational product generation, management, and constituent communities; and developing adaptive strategies for accommodating climate variability, uncertainty, and change. Other end-to-end efforts focus on the development of decision-support tools (DST) that embody unique resource management circumstances to enable formal and more objective linkages between meteorological, hydrologic, and institutional processes. Typically, end-to-end DST applications are developed for organizations making decisions with high-impact (e.g., state or national agencies) or high-economic value (e.g., hydropower production) and possess the technical and managerial abilities to efficiently exploit research advances (e.g., Georgakakos et al., 1998, 2004, 2005; Georgakakos, 2006). If linked to socioeconomic models incorporating

detailed information about the choices open to decision makers and their tolerance for risk. These end-to-end tools could also enable explicit assessment of the impacts of scientific and technological research advances.

This chapter describes a river management DST, RiverWare, which facilitates coordinated efforts among the research, operational product generation, and water management communities. RiverWare emerged from an early and sustained effort by several federal agencies to develop generic tools to support the assessment of water resources management options in river basins with multiple reservoirs and multiple management objectives (Frevert et al., 2006). RiverWare was selected for use as a case study because it has been used in a variety of settings, by multiple agencies, over a longer period than many other water management DSTs. Furthermore, RiverWare can explicitly accommodate a broad range of resource management concerns (e.g., flood control, recreation, navigation, water supply, water quality, and power production). RiverWare can also consider perspectives ranging from day-to-day scheduling of operations to long-range planning and can accommodate a variety of climate observations, forecasts, and even climate change projections. RiverWare can incorporate hydrologic risk, whereby event consequences and their magnitudes are mediated by their probability of occurrence, in strategic planning applications and design studies, which can offer a way forward for decision makers reluctant to shift away from use of traditional, stationarity-based, statistical analysis of historical data (Lee, 1999; Davis and Pangburn, 1999).

2. Description of RiverWare

RiverWare is a software framework used to develop detailed models of how water moves and is managed throughout complex river basin systems. RiverWare applications include physical processes (e.g., streamflow, bank storage, and solute transport), infrastructure (e.g., reservoirs, hydropower generating turbines, spillways, and diversion connections), and policies (e.g., minimum instream flow requirements and trades between water users) (Zagona et al., 2001, 2005). At a minimum, RiverWare applications require streamflow hydrographs as input for multiple locations throughout a river system. While hydrographs can be generated within the DST, they can also be input from other sources, with the latter approach being especially important in advanced end-to-end assessments. Detailed discussion of the role of observations and considerations of global change using RiverWare are discussed in later sections. RiverWare can be applied to address diverse water management concerns, including real-time operations, strategic planning for seasonal to interannual variability in water supplies and demands, and examining the impacts of hydrologic non-stationarity. Because infrastructure, management rules and policies can be easily changed, RiverWare also allows examination of alternative options for achieving management objectives over short-, medium-, and long-term planning horizons.

RiverWare was developed by the University of Colorado-Boulder's Center for Advanced Decision Support for Water and Environmental Systems (CADSWES) in collaboration with the Bureau of Reclamation, Tennessee Valley Authority, and the Army Corps of Engineers (Frevert et al., 2006). CADSWES continues to develop and maintain the RiverWare software and offers training and support for RiverWare users (http://cadswes.colorado.edu). According to CADSWES, RiverWare is used by more than 75 federal and state agencies, private sector consultants, universities and research institutes, and water districts, among others.

Example Applications

Consistent with the intent of its original design, the use of RiverWare varies widely depending on the specific application. An early application was its use for scheduling reservoir operations by the Tennessee Valley Authority (Eshenbach et al., 2001). In that application, RiverWare was used to define the physical and economic characteristics of the multi-reservoir system, including power production economics, to prioritize the policy goals that governed the reservoir operations and to specify parameters for linear optimization of system objectives. In another application, RiverWare was used to balance the competing priorities of minimum instream flows and consumptive water use in the operation of the Flaming Gorge Reservoir in Colorado (Wheeler et al., 2002).

While day-to-day scheduling of reservoir operations is more a function of weather than climate, the use of seasonal climate forecasts to optimize reservoir operations has long been a goal for water resources management. RiverWare is being implemented for the Truckee-Carson River basin in Nevada to investigate the impact of incorporating climate outlooks into an operational water management framework that prioritizes irrigation water supplies, interbasin diversions, and fish habitat (Grantz et al., 2007). Another example application to the Truckee-Carson River using a hypothetical operating policy indicated that fish populations could benefit from purchases of water rights for reservoir releases to mitigate warm summer stream temperatures resulting from low flows and high air temperatures (Neumann et al., 2006).

RiverWare has also been used to evaluate politically charged management strategies, including water transfers proposed in California's Quantification Settlement Agreement and the Bureau of Reclamation's Inadvertent Overrun Policy, maintaining instream flows sufficient to restore biodiversity in the Colorado River delta and conserving riparian habitat while accommodating future water and power development in the Bureau of Reclamation Multiple Species Conservation Program (Wheeler et al., 2002). RiverWare also played a key role in negotiations by seven western states concerning how the Colorado River should be managed and the river flow should be distributed among the states during times of drought. The Bureau of Reclamation implemented a special version of the RiverWare model of the Colorado River and its many reservoirs, diversions, and watersheds (Jerla, 2005). The model was used to provide support to the Basin States Modeling Work Group Committee over an 18-month period as they assessed different operational strategies under different hydrologic scenarios, including extreme drought (U.S. Department of Interior, 2007).

Implementation

RiverWare requirements are multi-dimensional. A specific river system and its infrastructure operating policies are defined by data files supplied to RiverWare. This allows incorporation of new basin features (e.g., reservoirs), operating policies, and hydroclimatic conditions without users having to write software code. Utilities within RiverWare enable users to automatically execute many simulations, including accessing external data or exporting results of model runs. Users can also write new modules that CADSWES can integrate into RiverWare for use in other applications. For example, in an application for the Pecos River in New Mexico, engineers developed new methods and software code for realistic downstream routing of summer monsoon-related flood waves (Boroughs and Zagona, 2002). RiverWare is implemented for use on Windows or Unix Solaris systems, as described in the requirements document (http://cadswes.colorado.edu/PDF/RiverWare/RecommendedMinimumSystemsRequirements.pdf). An extensive manual is also available (http://cadswes.colorado.edu/PDF/RiverWare/documentation/).

RiverWare applications can be implemented by any group that can pay for access, both in terms of finances and educational effort. Development of RiverWare applications requires a site license from CADSWES. Significant investment is required to learn to use RiverWare as well. CADSWES offers two 3-day RiverWare training courses—an initial class covering general simulation modeling, managing scenarios, and incorporating policy options through rule-based simulation and a second class covering rule-based simulation in more detail, creating basin policies, and examining water policy options. Costs for the original license, annual renewals, technical support, and training require several thousand dollars. The costs of licensing and learning RiverWare mean that small communities and civic groups are unlikely to implement their own applications for assessing water management options. Rather, large agencies with technical staff or the financial means to fund university research or consultants are the most frequent users of RiverWare. The agencies then mediate the access of stakeholders to assessments of water management options through traditional public processes (e.g., U.S. Department of Interior, 2007). Conflicts may arise in having academic research groups conduct analyses funded by stakeholder groups, with inherent tensions between the open publication of research required by academia and the limited access to results required by strategic negotiations among interest groups.

3. Current and Future Use of Observations

The specific combination of observations used by a RiverWare application depends on both the decision context and the use of other models and DSTs to provide input to RiverWare that more comprehensively or accurately describe the character, conditions, and response of the river basin system. Figure 5-1 illustrates the information flow linking observations, RiverWare, other models and DSTs, and water management decisions; it shows that RiverWare has tremendous flexibility in the kinds of observations that could be useful in hydrologic modeling and river system assessment and management. The types of observations that may ultimately feed into RiverWare applications also depend on the time scale of the situation.

A detailed discussion of the role of satellite observations in RiverWare applications and selected input models and DSTs (e.g., the Bureau of Reclamation's ET Toolbox and PRMS) is given by the "Evaluation Report for AWARDS ET Toolbox and RiverWare Decision-Support Tools" (Hydrological Sciences Branch, 2007). Briefly, RiverWare can use a combination of observations from multiple sources, including satellites, products derived from land-atmosphere or hydrologic models, and combinations of both. Satellite observations can assist models in estimating ET, precipitation, snow water equivalent, soil moisture, groundwater storage and aquifer volumes, reservoir storage, and water quality, among other variables. Measurements from sensors aboard a variety of satellites are being considered for their usefulness within DST contexts and their impacts on reducing water management uncertainty, including

the Moderate Resolution Imaging Spectroradiometer (MODIS) sensor aboard the Earth Observing System (EOS) Terra and Aqua satellites, Landsat telemetry data, ASTER, Shuttle Radar Topography Mission (SRTM) , Advanced Microwave Scanning Radiometer–EOS, GRACE, CloudSat, Tropical Rainfall Mapping Mission, and others. Future and planned satellites with hydrologically relevant sensors and measurements include the Global Precipitation Mission and the NPOESS. Use of these observations can be enhanced by assimilating them into land surface models to produce spatially-distributed estimates of snowpack, soil moisture, evapotranspiration, energy fluxes, and runoff, which then provide inputs to RiverWare to base a more comprehensive assessment of river basin conditions. The land surface models include the Community Land Model, MOSAIC, Noah, and VIC, among others, supported by NASA's Land Data Assimilation System and Land Information System (NASA, 2006a).

NASA has several pilot projects specifically focused on assessing the impact of satellite observations in a variety of hydrologic models and DSTs as they feed into RiverWare applications (NASA, 2005, 2006b, 2007). For example, one project is comparing Terra and Aqua MODIS snow cover products for the Yakima-Columbia River basins with land-based snow telemetry measurements, testing their use for Land Information System simulations that also use the North American Land Data Assimilation System, connecting assimilated snow data with the MMS PRMS, and then supplying the simulated runoff as inputs to RiverWare. Another project on the Rio Grande River basin is assessing MODIS and Landsat data to improve evapotranspiration estimates generated by the Bureau of Reclamation DST, the AWARDS ET Toolbox, which provides water-demand time series to RiverWare. While application of specific hydrologic models and observations depend on the specific RiverWare application, significant processing of both model and observations are required and can be resource intensive (e.g., calibration and aggregation/disaggregation).

Operational scheduling of reservoir releases depend on orders of water from downstream users (e.g., irrigation districts) that are largely affected by day-to-day weather conditions as well as seasonally varying demands. In these cases, the important observations are the near real-time estimates of conditions within the river basin system (e.g., soil moisture or infiltration capacity), which affect the transformation of precipitation into runoff in the river system, relative to constraints on system operation (e.g., reservoir storage levels or water temperatures at specific river locations). Prospective meteorological impacts are buffered by those placing the water orders

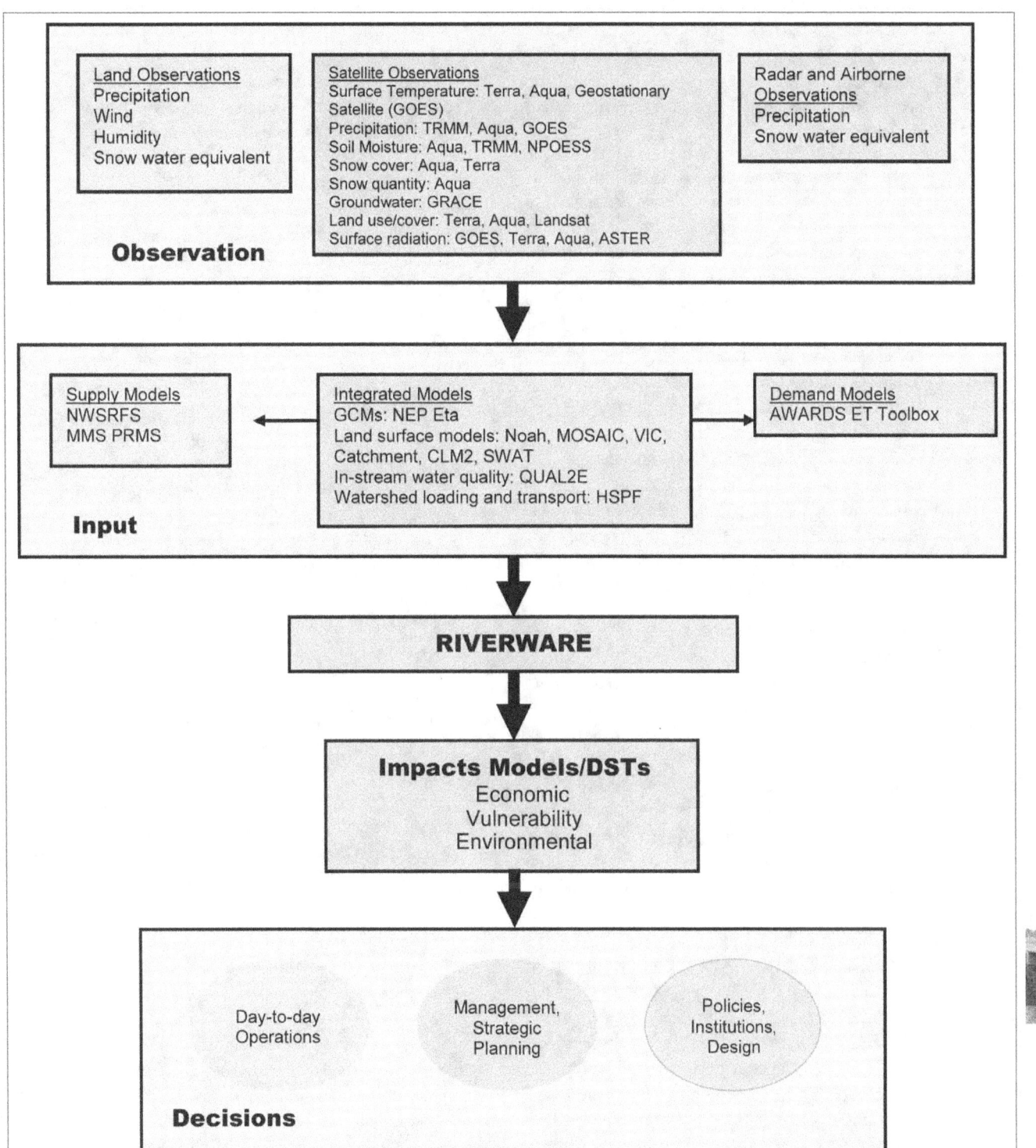

ASTER = Advanced Space-borne Thermal Emission and Reflection Radiometer; AWARDS = Agricultural Water Resources Decision Support; CLM2 = ; ET =
evapotranspiration; GCM = Global Climate Model; GRACE = Gravity Recovery and Climate Experiment; HSPF = Hydrological Simulation Program – Fortran;
Landsat = land remote-sensing satellite; MMS = Modular Modeling System; MOSAIC = ; NEP = Net Ecosystem Productivity; NOPESS = National Polar-Orbiting
Operational Environmental Satellite; NWSRFS = ; PRMS = Precipitation Runoff Modeling System; QUAL2E; SWAT = Soil and Water Assessment Tool; TRMM = ; VIC
= Variable Infiltration Capacity

Figure 5-1 Illustration depicting the flow of information.

or adjusting operations when the system is near some constraint (e.g., flood flows when reservoir levels are near peak storage capacity). In these situations, the important observations are recent extreme precipitation events and their location, which may be provided, separately or in some combination, by in-situ monitoring networks, radar, or satellites.

For mid-range applications, such as strategic planning for operations over the next season or year, outlooks of total seasonal water supplies are routinely used in making commitments for water deliveries, determining industrial and agricultural water allocation, and carrying out reservoir operations. In these applications, it is also important for water managers to keep track of the current state of the watershed. Such observations are often used as input to one of the many independent hydrologic models that can provide input to a specific RiverWare application. In these situations, the important observations are those that provide boundary or forcing conditions for the independent hydrologic models, including snowpack moisture storage, soil moisture, precipitation (intensity, duration, and spatial distribution), air temperature, humidity, winds, and other meteorological conditions.

For long-term planning and design applications, future meteorological uncertainty has a larger impact on outcomes than recent conditions based on observations; institutional change at multi-decadal time scales may have even greater impact. In these applications, accurate representation of anticipated natural hydroclimatological variability is important. In many western U.S. applications, observed streamflows are adjusted to remove the effects of reservoir management, interbasin diversions, and water withdrawals. The adjusted flows, termed "naturalized flows," may be used as input to RiverWare applications to assess the impact of different management options. Use of naturalized flows is fraught with problems. A central issue is poor monitoring of actual human impacts, especially withdrawals, diversions, and return flows (e.g., from irrigation). Alternative approaches include the use of proxy streamflows (e.g., from paleoclimatological indicators) or output from hydrologic modeling studies (Hartmann, 2005). For example, Tarboton (1995) developed hydrologic scenarios for severe sustained drought in the Colorado River basin based on streamflows reconstructed from centuries of tree-ring records; the scenarios were used in an assessment of management options using a precursor to the current RiverWare application to the Colorado River system.

The usefulness of the observations used within RiverWare depends on the specific implementation, as well as

the quality of the information itself. For example, one direct use of climate information for long-term planning includes hydrologic and hydraulic routing of "design storms" of various magnitudes and likelihoods, with the storms based on analyses of the available instrumental record (Urbanas and Roesner, 1993). However, those instrumental records have often been too short to adequately express climate variability and resulting impacts, regardless of the specific DSTs used to do the hydrologic or hydraulic routing. In short- and mid-range forecasting applications, the use of observations is mediated by the hydrologic model or DST that transforms weather and climate into streamflows, evaporative water demands, and other hydrologic processes. In these situations, from an operational perspective, the stream of observational inputs must be dependable, without downtime or large data gaps, and data processing, model simulation, and creation of forecast products must be fast and efficient. The usefulness of observations may be limited by other issues as well. The water resources management milieu is complex and diverse, and climate influences are only one factor among many affecting water management policies and practices. Factors limiting the use of observations or subsequent hydrologic model input to RiverWare for actual water management include lack of familiarity with the available information, disconnects between the specific information available (e.g., variables and spatiotemporal scales) and their relevance to decision makers, skepticism about the quality and applicability of information, conservative decision preferences due to accountability for poor consequences, and institutional impediments such as the inflexible nature of many multi-jurisdictional water management agreements (Changnon, 1990; Kenney, 1995; Pulwarty and Redmond, 1997; Pagano et al., 2001, 2002; Jacobs, 2002; Jacobs and Pulwarty, 2003; Rayner et al., 2005).

4. Uncertainty

The reliability of observations for driving hydrologic models that may provide input to RiverWare applications is the subject of much ongoing research. The hydrologic models, because they incompletely describe the physical relationships among important watershed components (e.g., vegetation processes that link the atmosphere and different levels of soil and surface and groundwater interactions), are themselves the subject of much research to determine their reliability. Streamflow and other hydrologic variables are intimately responsive to atmospheric factors, especially precipitation, that drive a watershed's behavior; however, errors in precipitation estimates are often amplified in the hydrologic response (Oudin et al., 2006).

Obtaining quality precipitation estimates is a formidable challenge, especially in the western U.S. where orographic effects produce large spatial variability and where there is a scarcity of real-time precipitation gage data and radar beam blockage by mountains. In principal, outputs from atmospheric models can serve as surrogates for observations, and provide forecasts of meteorological variables that can be used to drive hydrologic models. One issue in integrating atmospheric model output into hydrologic models for small watersheds (<1,000 km2) is that the spatial resolution of atmospheric models is lower than the resolution of hydrologic models. For example, quantitative precipitation forecasts produced by some atmospheric models may cover several thousand square kilometers, but the hydrologic models used for predicting daily streamflows require precipitation to be downscaled to precipitation fields for watersheds covering only tens or hundreds of square kilometers. One approach to produce output consistent with the needs of hydrologic models is to use nested atmospheric models, whereby outputs from large scale but coarse resolution models are used as boundary conditions for models operating over smaller domains with higher resolution. However, the error characteristics of atmospheric model products (e.g., bias in precipitation and air temperature) also can have significant effects on subsequent streamflow forecasts. Bias corrections require knowledge of the climatologies (i.e., long-term distributions) of both modeled and observed variables.

Although meteorological uncertainty may be high for the periods addressed by streamflow forecasts, accurate estimates of the state of watershed conditions prior to the forecast period are important because they are used to initialize hydrologic model states with significant consequences for forecast results. However, watershed conditions can be difficult to measure, especially when streamflow forecasts must be made quickly, as in the case of flash-flood forecasts. One option is to continuously update watershed states by running the hydrologic models continuously and by using inputs from recent meteorological observations and/or atmospheric models. Regardless of the source of inputs, Westrick et al. (2002) found it essential to obtain observational estimates of initial conditions to keep streamflow forecasts realistic; storm-by-storm corrections of model biases determined over extended simulation periods were insufficient. Recent experimental end-to-end forecasts of streamflow produced in a simulated operational setting (Wood et al., 2001) highlighted the critical role of quality estimates of spring and summer soil moisture used to initialize hydrologic model states for the eastern U.S.

Where streamflows may be largely comprised of snowmelt runoff, quality estimates of snow conditions are important. The importance of reducing errors in the timing and magnitude of snowmelt runoff are especially acute in regions where a large percentage of annual water supplies derive from snowmelt runoff, where snowmelt impacts are highly non-linear with increasing deviation from long-term average supplies, and where reservoir storage is smaller than interannual variation of water supplies. However, resources for on-site monitoring of snow conditions have diminished rather than grown relative to the increasing costs of errors in hydrologic forecasts (Davis and Pangburn, 1999). Research activities of the NWS National Office of Hydrology Remote-Sensing Center have long been directed at improving estimates of snowpack conditions through aerial and satellite remote sensing (Carroll, 1985). However, the cost of aerial flights prohibits routine use (T. Carroll, National Office of Hydrology Remote-Sensing Center, personal communication, 1999), while satellite estimates have qualitative limitations (e.g., not considering fractional snow coverage over large regions) and have not found broad use operationally.

Multiple techniques exist to more accurately represent the uncertainty inherent in understanding and predicting potential hydroclimatic variability. Stochastic hydrology techniques use various forms of autoregressive models to generate multiple synthetic streamflow time series with statistical characteristics matching available observations. For example, in estimating the risk of low flows for the Sacramento River Basin in California, the Bureau of Reclamation (Frevert et al., 1989) generated twenty 1,000-year streamflow time series matching selected statistics of observed flows (adjusted to compensate for water management impacts on natural flows); the non-exceedance probabilities of low flows were computed by counting the occurrences of low flows within 1- through 10-year intervals for all twenty 1,000-year sequences. The U.S. Army Corps of Engineers (1992) used a similar approach to estimate flood magnitudes with return periods exceeding 1,000 years using Monte Carlo sampling from within the 95% confidence limits of a Log Pearson III distribution developed by synthesizing multiple streamflow time series.

The capability to automatically execute many model runs within RiverWare, including accessing data from external sources and exporting model results, facilitates using stochastic hydrology approaches for representing uncertainty. For example, Carron et al. (2006) demonstrated RiverWare's capability to identify and quantify significant sources of uncertainty in projecting river and reservoir conditions using a first-order, second-moment algorithm that is computationally more efficient than more traditional Monte Carlo approaches. The first-order, second-moment processes uncertainties in inputs and models to provide estimates of uncertainty in model results that can be used directly within a risk management decision framework. The case study presented by Carron et al. (2006) evaluated the uncertainties associated with meeting goals for reservoir water levels beneficial for recovering endangered fish species within the lower Colorado River.

With regard to RiverWare applications concerned with mid-range planning and use of hydrologic forecasts, at the core of any forecasting system is the predictive model, whether a simple statistical relationship or a complex dynamic numerical model. Advances in hydrologic modeling have been notable, especially those associated with the proper identification of a model's parameters (e.g., Duan et al., 2002) and the development of models that consider the spatially distributed characteristics of watersheds rather than treating entire basins as a single point (Grayson and Bloschl, 2000). Conceptual rainfall-runoff models offer some advantages over statistical techniques in support of long-range planning for water resources management. These models represent, with varying levels of complexity, the transformation of precipitation and other meteorological forcing variables (e.g., air temperature and humidity) to watershed runoff and streamflow, including accounting for hydrologic storage conditions (e.g., snowpack, soil moisture, and groundwater). These models can be used to assess the impacts and implications of various climate scenarios by using historic meteorological time series as input, generating hydrologic time series, and then using those hydrologic scenarios as input to RiverWare. This approach enables consideration of current landscape and river channel conditions, which may be quite different than recorded in early instrumental records and can dramatically alter a watershed's hydrologic behavior (Vorosmarty et al., 2004). Furthermore, the use of multiple input time series, system parameterizations, or multiple models enables a probabilistic assessment of an ensemble of scenarios. The Hydrological Ensemble Prediction Experiment (Schaake et al., 2007) aims to address the unique challenges of expressing uncertainty associated with ensemble forecasts for water resources management.

An additional concern for mid- and long-range planning is that, as instrumental records have grown longer, they often show trends (e.g., Baldwin and Lall, 1999; Olsen et al., 1999; Andreadis and Lettenmaier, 2006) or persistent regimes (i.e., periods characterized by distinctly different statistics) (e.g., Angel and Huff, 1995; Quinn, 1981, 2002), with consequences for estimation of hydrologic risk (Olsen et al., 1998). Observed regimes and trends can have multiple causes, including climatic changes, watershed and river transformations,

and management impacts (e.g., irrigation return flows and transbasin water diversions). These issues enter into RiverWare applications directly through the use of naturalized flows, which are notoriously unreliable. For example, in assessments of water management options on the San Juan River in Colorado and New Mexico, the reliability of naturalized flows was considered to be affected by the inconsistent accounting of consumptive uses between irrigation and non-irrigation data, use of reservoir evaporation rates with no year-to-year variation, neglecting time lags in the accounting of return flows from irrigation to the river, errors in river gage readings that underestimated flows in critical months, and the lack of documentation of diversions that reduce river flows as well as subsequent adjustments to data used to compute naturalized flows.

5. Global Change Information and RiverWare

Climate Variability

Decision makers increasingly recognize that climate is an important source of uncertainty and potential vulnerability in long-term planning for the sustainability of water resources (Hartmann, 2005). With the appropriate investment in site licenses, training of personnel, implementation for a specific river system, and assessment efforts, RiverWare is capable of supporting climate-related water resources management decisions by U.S. agencies. However, technology alone is insufficient to resolve conflicts among competing water uses. Early in the development of RiverWare, Reitsma et al. (1996) investigated its potential role as a DST within complex negotiations between hydroelectric, agricultural, and flood control interests. Results indicated that while DSTs can help identify policies that can satisfy specific management requirements and constraints, as well as expand the range of policy options considered, they are of limited value in helping decision makers understand interactions within the river system. Furthermore, the burdens of direct use by decision makers of a DST that embodies a complex system are significant; a more useful approach is to have specialists support decision makers by making model runs and presenting the results in an iterative manner. This is the approach used by the Bureau of Reclamation in the application of RiverWare to support interstate negotiations concerning the sharing of Colorado River water supply shortages during times of drought (Jerla, 2005; U.S. Department of Interior, 2007).

From the perspective of mid-range water management issues, the use of forecasts within RiverWare applications constitutes an important pathway for supporting climate-related decision making. Each time a prediction is made, science has an opportunity to address and communicate the strengths and limitations of current understanding. Each time a decision is made, managers have an opportunity to confront their understanding of scientific information and forecast products. Furthermore, each prediction and decision provides opportunities for interaction between scientists and decision makers and for making clear the importance of investments in scientific research. Perceptions of poor forecast quality are a significant barrier to more effective use of hydroclimatic forecasts (Changnon, 1990; Pagano et al., 2001, 2002; Rayner et al., 2005); however, recent advances in modeling and predictive capabilities naturally lead to speculation that hydroclimatic forecasts can be used to improve the operation of water resource systems.

Great strides have been made in monitoring, understanding, and predicting interannual climate phenomena such as the El Nino-Southern Oscillation (ENSO). This improved understanding has resulted in long-lead (up to about a year) climate forecast capabilities that can be exploited in streamflow forecasting. Techniques have been developed to directly incorporate variable climate states into probabilistic streamflow forecast models based on linear discriminant analysis with various ENSO indicators, (e.g., the Southern Oscillation Index) (Peichota and Dracup, 1999; Piechota et al., 2001). Recent improved understanding of decadal-scale climate variability also has contributed to improved interannual hydroclimatic forecast capabilities. For example, the Pacific Decadal Oscillation (PDO) (Mantua et al., 1997) has been shown to modulate ENSO-related climate signals in the West. Experimental streamflow forecasting systems for the Pacific Northwest have been developed based on long-range forecasts of both PDO and ENSO (Hamlet and Lettenmaier, 1999). In the U.S., the Pacific Northwest, California, and the Southwest are strong candidates for the use of long-lead forecasts because ENSO and PDO signals are particularly strong in these regions, and each region's water supplies are closely tied to accumulation of winter snowfall, amplifying the impacts of climatic variability.

While many current water management decision processes use single-value deterministic approaches, probabilistic forecasts enable quantitative estimation of the inevitable uncertainties associated with weather and climate systems. From a decision maker's perspective, probabilistic forecasts are more informative because they explicitly communicate uncertainty and are more useful because they can be directly incorporated into risk-based calculations. Probabilistic forecasts of water supplies can be created by overlaying a single prediction with a normal distribution of estimation error determined at the time of calibration of the forecast

equations (Garen, 1992). However, to account for future meteorological uncertainty, new developments have focused on ensembles, whereby multiple possible futures (each termed an ensemble trace) are generated; statistical analysis of the ensemble distribution then provides the basis for a probabilistic forecast.

Changnon (2000), Rayner et al. (2005), and Pagano et al. (2002) found that improved climate prediction capabilities are initially incorporated into water management decisions informally using subjective, ad-hoc procedures on the initiative of individual water managers. While improvised, those decisions are not necessarily insignificant. For example, the Salt River Project, among the largest water management agencies in the Colorado River Basin and primary supplier to the Phoenix metropolitan area, decided in August 1997 to substitute groundwater withdrawals with reservoir releases, expecting increased surface runoff during a wet winter related to El Nino. With that decision, they risked losses exceeding $4 million in an attempt to realize benefits of $1 million (Pagano et al., 2002). Because these informal processes are based in part on confidence in the predictions, overconfidence in forecasts can be even more problematic than lack of confidence as a single incorrect forecast that provokes costly shifts in operations can devastate user confidence in subsequent forecasts (e.g., Glantz, 1982).

The lack of verification of hydroclimatic forecasts is a significant barrier to their application in water management, but it is not easy to resolve with traditional research efforts because the level of acceptable skill varies widely depending on the intended use (Hartmann et al., 2002a; Pagano et al., 2002). Information on forecast performance has rarely been available to and framed for decision makers, although hydrologic forecasts are reviewed annually by the issuing agencies in the U.S (Hartmann et al., 2002b). Hydrologic forecast verification is an expanding area of research (Franz et al., 2003; Hartmann et al., 2003; Bradley et al., 2004; Pagano et al., 2004; Kruger et al., 2007), but much work remains and could benefit from approaches developed within the meteorological community (Welles et al., 2007). Because uncertainty exists in all phases of the forecast process, forecast systems designed to support risk-based decision making need to explicitly quantify and communicate uncertainties from the entire forecast system and from each component source, including model parameterization and initialization, meteorological forecast uncertainty at the multiple spatial and temporal scales at which they are issued, adjustment of meteorological forecasts (e.g., through downscaling) to make them usable for hydrologic models, implementation of ensemble techniques, and verification of hydrologic forecasts.

Climate Change

From the perspective of long-range water management issues, the potential impacts of climate change on water resources and their implications for management are central topics of concern. Estimates of prospective impacts of climate change on precipitation have been mixed, leading, in many cases, to increasing uncertainty about the reliability of future water supplies. However, where snow provides a large fraction of annual water supplies, prospective temperature increases dominate hydrologic impacts, leading to stresses on water resources and increased hydrologic risk. Higher temperatures effectively shift the timing of the release of water stored in the snowpack "reservoir" to earlier in the year, reducing supplies in summer when demands are greatest, while also increasing the risk of floods due to rain-on-snow events. While not using RiverWare, several river basin studies have assessed the risks of higher temperatures on water supplies and management challenges. The near universal analytical approach has been one of sensitivity analysis (Lettenmaier, 2003):

1. Downscaling outputs from a dynamic general circulation model of the global land-atmosphere-ocean system to generate regional- or local-scale meteorological time series over many decades,

2. Using the meteorological time series as input to rainfall-runoff models to generate hydrologic time series,

3. Using the hydrologic scenarios as input to water management models, and

4. Assessing differences among baseline and change scenarios using a variety of metrics.

Early assessments of warming impacts on large river basins generally showed extant water management systems to be effective for all but the most severe scenarios (Hamlet and Lettenmaier, 1999; Lettenmaier et al., 1999), with a notable exception being the Great Lakes system, where increased lake heat storage was tied to loss of ice cover, increased winter lake evaporation, lower lake levels, and potential failure to meet Lake Ontario regulation objectives under extant operating rules (Croley, 1990; Hartmann, 1990; Lee et al., 1994; Lee et al., 1997; Sousounis et al., 2000; Lofgren et al., 2002).

Extensive detailed studies of the capability of existing reservoir systems and operational regulation rules to meet water management goals under changed climates are fairly recent (e.g., Saunders and Lewis, 2003; Christensen, et al., 2004; Payne et al., 2004; VanRheenan et al., 2004; Maurer, 2007). However, there is a rapidly growing literature on broad considerations of climate change in water resources management (Frederick

et al., 1997; Gamble et al., 2003; Lettenmaier, 2003; Loomis et al., 2003; Snover et al., 2003; Stakhiv, 2003; Ward et al., 2003; Vicuna et al., 2007). Some (Matalas, 1997) that contend that existing approaches are sufficient for water resource management planning and risk assessment because they contain safety factors; however, an inescapable message for the water resource management community is the inappropriateness of the stationarity assumption in the face of climate change. While precipitation changes may remain too uncertain for consideration in the near term, temperature increases are more certain and can have strong hydrologic consequences.

Cognitively, climate change information is difficult to integrate into water resources management. First, within the water resources engineering community, the stationarity assumption is a fundamental element of professional training. Second, the century time scales of climate change exceed typical planning and infrastructure design horizons and are remote from human experience. Third, even individuals trying to stay up-to-date can face confusion in conceptually melding the burgeoning climate change impacts literature. Assessments are often repeated as general circulation and hydrologic model formulations advance or as new models become available throughout the research community. Furthermore, assessments can employ a variety of techniques for downscaling. Transposition techniques (e.g., Croley et al., 1998) are more intuitive than the often mathematically complex statistical and dynamical downscaling techniques (e.g., Clark et al., 1999; Westrick and Mass, 2001; Wood et al., 2002; Benestad, 2004).

GCMs and their downscaled corollaries provide one unique perspective on long-term trends related to global change. Another unique perspective is provided by tree-ring reconstructions of paleo-streamflows, which, for example, indicate that in the U.S. Southwest droughts over the past several hundred years have been more intense, regionally extensive, and persistent than those reflected in the instrumental record (Woodhouse and Lukas, 2006). Decision makers have expressed interest in combining the perspectives of paleoclimatological information and GCMs. While some studies have linked instrumental records to paleoclimatological information (e.g., Prairie, 2006) and others with GCMs (e.g., Christensen and Lettenmaier, 2006), few link all three (an exception is Smith et al., 2007).

Conceptual integration of climate change impacts assessment results in a practical water management context is complicated by the multiplicity of scenarios and vague attribution of their prospects for occurrence, which depend so strongly on feedbacks among social,

economic, political, technological, and physical
processes. For decision makers, a critical issue concerns
the extent to which the various scenarios reflect the
actual uncertainty of the relevant risks versus the
uncertainty due to methodological approaches and biases
in underlying models. The difficulties facing decision
makers in reconciling disparate climate change impact
assessments are exemplified by the Upper Colorado
River Basin, where reductions in naturalized flow by
the mid-21st century have been estimated to range from
about 45% by Hoerling and Eischeid (2007), 10 to 25%
by Milly et al (2005), about 18% by Christensen et al.
(2004), and about 6% by Christensen and Lettenmaier
(2006). Furthermore, using the difference between
precipitation and evapotranspiration as a proxy for runoff,
Seager et al. (2007) suggest an "imminent transition to a
more arid climate in southwestern North America."

However, in the face of circumstances nearing or
exceeding the effectiveness of existing management
paradigms, individuals can become more cognizant of the
need to consider climate change. In the U.S. Southwest
between 1999 and 2004, Lake Powell levels declined
faster than previously considered in scenarios of extreme
sustained drought (e.g., Harding et al., 1995; Tarboton,
1995), from full to only 38% capacity in November 2004
(Bureau of Reclamation, 2004). Resource managers,
policymakers, and the general public are now actively
seeking scientific guidance in exploring how management
practices can be more responsive to the uncertainties
associated with a changing climate.

APPENDIX A

Chapter I References – Decision Support for Agricultural Efficiency

Birkett, Charon and Brad Doorn. 2004. "A New Remote Sensing Tool for Water Resources Management, *Earth Observation Magazine*, October 13 (6).

Congressional Research Service, Science Policy Research Division. 1983. "United States Civilian Space Programs, Volume II Applications Satellites," prepared for the Subcommittee on Space Science and Applications of the Committee on Science and Technology, U.S. House of Representatives, May. FAS OnLine Crop Assessment at http://www.fas.usda.gov/pecad2/crop_assmnt.html (accessed April 2007).

Kanarek, Harold. 2005. "The FAS Crop Explorer: A Web Success Story," *FAS Worldwide*, June (http://www.fas.usda.gov/info/fasworldwide/2005/06-2005/Cropexplorer.htm (accessed April 2007).

Kaupp, Verne, Charles Hutchinson, Sam Drake, Tim Haithcoat, Willem van Leeuwen, Vlad Likholetov, David Tralli, Rodney McKellip, and Brad Doorn. 2005. "Benchmarking the USDA Production Estimates and Crop Assessment Division DSS Assimilation," v.3 (01.04.06), report prepared for Production Estimates and Crop Assessment Division, FAS, U.S. Department of Agriculture, September.

National Aeronautics and Space Administration, 2001. Aeronautics and Space Report of the President, NASA, Washington DC at http://history.nasa.gov/presrep01/pages/usda.html (accessed April 2007).

National Aeronautics and Space Administration, John C. Stennis Space Center. 2004a. "Decision-support Tools Evaluation Report for FAS/PECAD," Version 2.0, January.

National Aeronautics and Space Administration, John C. Stennis Space Center. 2004b. "PECAD's Global Reservoir and Lake Monitor: A Systems Engineering Report," Version 1.0, December.

National Aeronautics and Space Administration. 2006a. "NASA Science Mission Directorate: Earth-Sun System Applied Sciences Program Agricultural Efficiency Program Element FY2006-2010 Plan,"30 June at http://aiwg.gsfc.nasa.gov/esappdocs/Agricultural_Efficiency_FINAL_06.pdf (accessed April 2007).

National Aeronautics and Space Administration, 2006b. "NASA Science Mission Directorate – Applied Sciences Program: Agricultural Efficiency – FY 2005 Annual Report at http://aiwg.gsfc.nasa.gov/esappdocs/annualreports/ (accessed April 2007).

National Assessment Synthesis Team. 2004. *Climate Change Impacts on the United States: The Potential Consequences of Climate Variability* (Boston, MA: Cambridge University Press).

National Research Council, Board on Earth Sciences and Resources. 2007. *Contributions of Land Remote Sensing for Decisions about Food Security and Human Health: Workshop Report* (Washington, DC: National Academies Press).

Reynolds, Curt A. 2001. "CADRE Soil Moisture and Crop Models," at http://www.pecad.fas.usda.gov/cropexplorer/datasources.cfm (accessed April 2007).

Rosenzweig, Cynthia. 2003. "Climate Change and Agriculture: Mitigation and Adaptation," Testimony before the Senate Committee on Environment and Public Works, Subcommittee on Clean Air, Climate Change, and Nuclear Safety, July 8 at http://epw.senate.gov/108th/Rosenzweig_070803.htm (accessed April 2007).

United Nations Food and Agriculture Organization. 2006. "Agricultural Monitoring Meeting Convened for the Integrated Global Observations for Land Theme," Rome, Italy (8–11 March 2006), 28 June.

United Nations Food and Agriculture Organization. No date. "Agriculture and Climate Change: FAO's Role" at http://www.fao.org/News/1997/971201-e.htm (accessed April 2007).

Chapter 2 References – Decision Support for Air Quality

Al-Saadi, J., J. Szykman, R. B. Pierce, C. Kittaka, D. Neil, D. A. Chu, L. Remer, L. Gumley, E. Prins, L. Weinstock, C. MacDonal, R. Wayland, F. Dimmick, and J. Fishman, 2005: Improving national air quality forecast with satellite aerosol observations. Bulletin of the American Meteorological Society, Volume 86, Issue 9, 1249–1261.

Bey, I., D. J. Jacob, R. M. Yantosca, J. A. Logan, B. D. Field, A. M. Fiore, Q. Li, H. Y. Liu, L. J. Mickley, and M. G. Schultz (2001), Global modeling of tropospheric chemistry with assimilated meteorology: Model description and evaluation, J. Geophys. Res., 106, 23,073–23,095.

Brasseur, G. P., J. T. Kiehl, J.-F. Mueller, T. Schneider, C. Granier, X. X. Tie, and D. Hauglustaine, 1998: Past and future changes in global tropospheric ozone: Impact on radiative forcing, Geophys. Res. Lett., 25, 3807–3810.

Brown, T.J., B.L. Hall, and A.L. Westerling, 2004: The impact of twenty-first century climate change on wildland fire danger in the western United States: An applications perspective. Climatic Change, 62, 365–388.

Byun, D.W., 1999a: Dynamically consistent formulations in meteorological and air quality models for multi-scale atmospheric applications: Part I. Governing equations in generalized coordinate system. Journal of Atmospheric Science, Vol 56, 3789–3807.

Byun, D.W., 1999b: Dynamically consistent formulations in meteorological and air quality models for multi-scale atmospheric applications: Part II. Mass conservation issues. Journal of Atmospheric Science, Vol 56, 3808–3820.

Byun, D.W. and Ching, J.K.S. (eds.), 1999: Science algorithms of the EPA Models-3 Community Multiscale Air Quality Model (CMAQ) modeling system. EPA/600/R-99/030, U.S. Environmental Protection Agency, Office of Research and Development, Washington, DC 20460.

Byun, D.W., and K. L. Schere, 2006: Review of the Governing Equations, Computational Algorithms, and Other Components of the Models-3 Community Multiscale Air Quality (CMAQ) Modeling System . Applied Mechanics Reviews, Volume 59, Number 2 (March 2006), pp. 51–77.

Carmichael, G. R., L. K. Peters, and R. D. Saylor, 1991: The STEM-II regional scale acid deposition and photochemical oxidant model—I. An overview of model development and applications, Atmos. Environ., 25(10), 2077–2090

Civerolo, K., C. Hogrefe, B. Lynn, J. Rosenthal, J.-Y. Ku, W. Solecki, J. Cox, C. Small, C. Rosenzweig, R. Goldberg, K. Knowlton, and P. Kinney, 2007: Estimating the effects of increased urbanization on surface meteorology and ozone concentrations in the New York City metropolitan region. Atmos. Environ., 41, 1803–1818, doi:10.1016/j.atmosenv.2006.10.076.

CMAS, 2007: Community Modeling and Analysis System. Available at http://www.cmascenter.org/.

Constantinescu, E.M., A. Sandu, T. Chai, and G.R. Carmichael, 2007a: Ensemble-based Chemical Data Assimilation I: General Approach. Quarterly Journal of the Royal Meteorological Society, Vol 133(626), Pages 1229–1243.

Constantinescu, E.M. , A. Sandu, T. Chai, and G.R. Carmichael, 2007b: Ensemble-based Chemical Data Assimilation II: Covariance Localization. Quarterly Journal of the Royal Meteorological Society, Vol 133(626), Pages 1245–1256.

Delworth, T.L., A. J. Broccoli, A. Rosati, R. J. Stouffer, V. Balaji, J. A. Beesley, W. F. Cooke, and 37 co-authors, 2006: GFDL's CM2 Global Coupled Climate Models. Part I: Formulation and Simulation Characteristics, Journal of Climate-Special Section, Vol. 19, 643–674

Duncan, B.N., R.V. Martin, A.C. Staudt, R. Yevich, J.A. Logan, 2003: Interannual and Seasonal Variability of Biomass Burning Emissions Constrained by Satellite Observations, J. Geophys. Res., 108(D2), 4040, doi:10.1029/2002JD002378.

Eder, B., D. Kang, R. Mathur, S. Yu, K. Schere, 2006: An operational evaluation of the Eta-CMAQ air quality forecast model, Atmospheric Environment 40, 4894–4905

Emmons, L. K., D. A. Hauglustaine, J.-F. Muller, M. A. Carroll, G. P. Brasseur, D. Brunner, J. Staehelin, V. Thouret, and A. Marenco, 2000: Data composites of airborne observations of tropospheric ozone and its precursors, J. Geophys. Res., 105, 20,497–20,538.

EPA, 1995. Appendix W to Part 51 of 40CFR: Guideline on Air Quality Models. Available at http://www.epa.gov/fedrgstr/EPA-AIR/1995/August/Day-09/pr-912.html.

EPA, 2007a: Community Multiscale Air Quality (CMAQ) modeling system. Available at http://www.epa.gov/asmdnerl/CMAQ/.

EPA, 2007b: Emissions Modeling Clearinghouse, Biogenic Emission Sources. Available at http://www.epa.gov/ttn/chief/emch/biogenic/.

Friedl, R. (ed.), 1997: Atmospheric effects of subsonic aircraft: Interim assessment report of the advanced subsonic technology program, NASA Ref. Publ. 1400, 143 pp., 1997.

Fu, T.-M., D. J. Jacob, P. I. Palmer, K. Chance, Y. X. Wang, B. Barletta, D. R. Blake, J. C. Stanton, and M. J. Pilling, 2007: Space-based formaldehyde measurements as constraints on volatile organic compound emissions in east and south Asia and implications for ozone, J. Geophys. Res., 112, D06312, doi:10.1029/2006JD007853.

Hakami, A., D.K. Henze, J.H. Seinfeld, K. Singh, A. Sandu, S. Kim, D. Byun, and Q. Li, 2007: The adjoint of CMAQ, (in print).

Hansen, J., Mki. Sato, L. Nazarenko, R. Ruedy, A. Lacis, D. Koch, I. Tegen, T. Hall, and 20 co-authors, 2002: Climate forcings in Goddard Institute for Space Studies SI2000 simulations. J. Geophys. Res. 107, no. D18, 4347, doi:10.1029/2001JD001143.

Hansen, J., Mki. Sato, R. Ruedy, L. Nazarenko, A. Lacis, G.A. Schmidt, G. Russell, and 38 co-authors, 2005: Efficacy of climate forcings. J. Geophys. Res. 110, D18104, doi:10.1029/2005JD005776.

Harvard, 2007: Harvard Atmospheric Chemistry Modeling Group. Available at http://www-as.harvard.edu/chemistry/trop.

Heald, Colette L., Daniel J. Jacob, Paul I. Palmer, Mathew J. Evans, Glen W. Sachse, Hanwant B. Singh and Donald R. Blake, 2003: Biomass burning emission inventory with daily resolution: application to aircraft observations of Asian outflow, J. Geophys. Res., 108(D4), 8368, doi:10.1029/2002JD002732.

Hoelzemann, J.J., M. G. Schultz, G. P. Brasseur, and C. Granier, 2004: Global Wildland Fire Emission Model (GWEM): Evaluating the use of global area burnt satellite data, J. Geophys. Res., 109, D14S04, doi:10.1029/2003JD003666.

Hogrefe, C., B. Lynn, K. Civerolo, J.-Y. Ku, J. Rosenthal, C. Rosenzweig, R. Goldberg, S. Gaffin, K. Knowlton, and P. L. Kinney, 2004: Simulating changes in regional air pollution over the eastern United States due to changes in global and regional climate and emissions, J. Geophys. Res., 109, D22301, doi:10.1029/2004JD004690.

Hogrefe C, LR Leung, LJ Mickley, SW Hunt, and DA Winner, 2005: Considering Climate Change in U.S. Air Quality Management. EM: Air & Waste Management Association's magazine for environmental managers October 2005:19–23.

Holloway, T., H. Levy II, and G. Carmichael, 2002: Transfer of reactive nitrogen in Asia: development and evaluation of a source-receptor model. Atmospheric Environment, 36(26), 4251–4264.

Horowitz, L. W., and Coauthors, 2003: A global simulation of tropospheric ozone and related tracers: Description and evaluation of MOZART, version 2. J. Geophys. Res., 108, 4784, doi:10.1029/2002JD002853.

Hurrell, J.W., J.J. Hack, A.S. Phillips, J. Caron, and J. Yin, 2006: The Dynamical Simulation of the Community Atmosphere Model Version 3 (CAM3) Journal of Climate: Vol. 19, pp 2162-2183.

In, H.-J., D. W. Byun, R. J. Park, N.-K. Moon, S. Kim, and S. Zhong, 2007: Impact of transboundary transport of carbonaceous aerosols on the regional air quality in the United States: A case study of the South American wildland fire of May 1998, J. Geophys. Res., 112, D07201, doi:10.1029/2006JD007544.

IPCC (Intergovernmental Panel on Climate Change), 2000: Emissions Scenarios, Cambridge University Press, Cambridge, UK

IPCC (Intergovernmental Panel on Climate Change), 2001: The Scientific Basis. Cambridge University Press, Cambridge, UK

Jacob, D.J., and A.B. Gilliland, 2005: Modeling the impact of air pollution on global climate change, Environmental Manager, pp. 24–27, October 2005, Air and Waste Management Association. Pittsburgh, PA.

Jacobson, M. Z., GATOR-GCMM, 2001a: A global through urban scale air pollution and weather forecast model. 1. Model design and treatment of subgrid soil, vegetation, roads, rooftops, water, sea ice, and snow. J. Geophys. Res., 106, 5385–5402.

Jacobson, M. Z., 2001b: GATOR-GCMM: 2. A study of day- and nighttime ozone layers aloft, ozone in national parks, and weather during the SARMAP Field Campaign, J. Geophys. Res., 106, 5403–5420, 2001

Kalkstein, L. S., and K. M. Valimont, 1987: Climate effects on human health. EPA Science and Advisory Committee Monograph, no. 25389: 122–152. Washington D. C., U. S. EPA.

Kiehl, J.T., J. Hack, G. Bonan, B. Boville, B. Briegleb, D. Williamson, and P. Rasch, 1996: Description of the NCAR Community Climate Model (CCM3). NCAR Technical Note. NCAR/TN-420+STR, Ntl. Center for Atmos. Research, Boulder, CO, 152 pp. [Available Ntl. Cen. Atmos. Res., P.O. Box 3000, Boulder, CO, 80305.]

Knowlton, K., Rosenthal, J.E., Hogrefe, C., Lynn, B., Gaffin, S., Goldberg, R., Rosenzweig, C., Civerolo, K., Ku, J.-Y., Kinney, P.L., 2004. Assessing ozone-related health impacts under a changing climate. Environ. Health Perspect. 2004, 112, 1557–1563.

Kopacz, M., D. J. Jacob, D. Henze, C. L. Heald, D. G. Streets, Q. Zhang, 2007:Comparison of adjoint and analytical Bayesian inversion methods for constraining Asian sources of carbon monoxide using satellite (MOPITT) measurements of CO columns, Journal of Geophysical Research – Atmospheres (in press).

Leung, L. R., and M. S. Wigmosta, 1999: Potential climate change impacts on mountain watersheds in the Pacific Northwest. J. Amer. Water Resour. Assoc., 35(6): 1463–1471.

Leung, L. R., S. J. Ghan, Z.-C. Zhao, Y. Luo, W.-C. Wang, and H. Wei, 1999: Intercomparison of regional climate simulations of the 1991 summer monsoon in East Asia. J. Geophys. Res., 104(D6): 6425–6454.

Leung LR, Y Kuo, and J Tribbia. 2006: Research Needs and Directions of Regional Climate Modeling Using WRF and CCSM." Bulletin of the American Meteorological Society 87(12):1747–1751.

Liang, X.-Z., J. Pan, J. Zhu, K.E. Kunkel, J.X.L. Wang, and A. Dai, 2006: Regional climate model downscaling of the U.S. summer climate and future change. J. Geophys. Res., 111, D10108.

Liao, K.-J., E. Tagaris, K. Manomaiphiboon, J.-H. Woo, S. He, P. Amar, and A.G. Russell, 2007: Sensitivities of Ozone and Fine Particulate Matter Formation to Emissions under the Impact of Potential Future Climate Change, Environmental Science & Technology (in print).

Logan, J.A., 1999: An analysis of ozonesonde data for the troposphere: Recommendations 601 for testing 3-D models and development of a gridded climatology for tropospheric ozone, J. 602 Geophys. Res., 104, D13, 16,115–16,149.

LRTAP, 2007a: Task Force on Hemispheric Transport of Air Pollution. Available at http://www.htap.org/index.htm.

LRTAP, 2007b: Task Force on Hemispheric Transport of Air Pollution, 2007 Interim Report. Available at http://www.htap.org/activities/2007_Interim_Report.htm.

Mearns, L.O., 2003: Issues in the impacts Climate variability and change on agriculture—Applications to the southeastern United States. Climate Change, 60, 1–6.

Mickley, L.J., D.J. Jacob, B.D. Field, and D. Rind, 2004: Effects of future climate change on regional air pollution episodes in the United States, Geophys. Res. Let., 30, L24103, doi:10.1029/2004GL021216.

NARSTO, 2000: An assessment of tropospheric ozone pollution-a North American Perspective. NARSTP Management Coordinator's Office (Envair). Pasco, Washington. Available at http://www.narsto.org/.

NASA, 2007a: CALIPSO, Cloud-Aerosol Lidar and Infrared Pathfinder Satellite Observations. Available at http://www-calipso.larc.nasa.gov/.

NASA, 2007b: Goddard Institute for Space Studies, Software Tools. http://www.giss.nasa.gov/tools/.

NASA, 2007c: Tropospheric Chemistry Integrated Data Center. Available at http://www-air.larc.nasa.gov/.

NOAA, 2007a Hazard Mapping System Fire and Smoke Product http://www.ssd.noaa.gov/PS/FIRE/hms.html.

NOAA, 2007b : Earth Systems Research Laboratory, Physical Science Division. Available at http://www.cdc.noaa.gov/cdc/data. ncep.reanalysis.html.

Novelli, P. C., K. A. Masarie, P. M. Lang, B. D. Hall, R. C. Myers, and J.W. Elkins, 2003: Reanalysis of tropospheric CO trends: Effects of the 1997–1998 wildfires, J. Geophys. Res., 108(D15), 4464, doi:10.1029/ 2002JD003031.

Pour-Biazar, A., R.T. McNider, S.J. Roselle, R. Suggs, G. Jedlovex, D.W. Byun, S.T. Kim, C.J. Lin, T.C. Ho, S. Haines, B. Dornblaser, and R. Cameron, 2007: Correcting photolysis rates on the basis of satellite observed clouds, J. Geophys. Res., 112, D10302, doi:10.1029/2006JD007422.

Russell, A., and R. Dennis, 2000: NARSTO critical review of photochemical models and modeling. Atmos. Environ., 34, 2283–2324.

Sandu, A., D. Daescu, G.R. Carmichael, and T. Chai, 2005: Adjoint sensitivity analysis of regional air quality models. Journal of Computational Physics, 204:222–252.

Schmidt, G.A., R. Ruedy, J.E. Hansen, I. Aleinov, N. Bell, M. Bauer, S. Bauer, B. Cairns, V. Canuto, Y. Cheng, A. Del Genio, G. Faluvegi, A.D. Friend, T.M. Hall, Y. Hu, M. Kelley, N.Y. Kiang, D. Koch, A.A. Lacis, J. Lerner, K.K. Lo, R.L. Miller, L. Nazarenko, V. Oinas, J. Perlwitz, D. Rind, A. Romanou, G.L. Russell, M. Sato, D.T. Shindell, P.H. Stone, S. Sun, N. Tausnev, D. Thresher, and M.S. Yao, 2006: Present day Atmos. simulations using GISS Model E: Comparison to in-situ, satellite and reanalysis data. J. Clim., 19, 153–192, doi:10.1175/JCLI3612.1.

Song, C.-K., D.W. Byun, R.B. Pierce, J.A. Alsaadi, T.K. Schaack, and F. Vukovich, 2007: Downscale linkage of global model output for regional chemical transport modeling: method and general performance. Journal of Geophysical Research. (in print)

Spak, S.N., T. Holloway, B. Lynn, R. Goldberg, 2007: A Comparison of Statistical and Dynamical Downscaling for Surface Temperature in North America, J. Geophys. Res., 112, D08101, doi:10.1029/2005JD006712.

Spracklen, D.V., J. A. Logan, L. J. Mickley, R. J. Park, R. Yevich, A.L. Westerling, and D. Jaffe, 2007: Wildfires drive interannual variability of organic carbon aerosol in the western U.S. in summer: implications for trends. Geophys. Res. Lett., 34, L16816, doi:10.0129/GL030037, 2007.

Tang, Y., G. R. Carmichael, N. Thongboonchoo, T. Chai, L. W. Horowitz, R. B. Pierce, J. A., Al-Saadi, G., Pfister, J. N. Vukovich, M. A. Avery, G. W. Sachse, T. B. Ryerson, J. S. Holloway, E. L. Atlas, F. M. Flocke, R. J. Weber, L. G. Huey, J. E. Dibb, D. G. Streets, W. H. Brune, 2007: Influence of lateral and top boundary conditions on regional air quality prediction: A multiscale study coupling regional and global chemical transport models, J. Geophys. Res., 112, D10S18, doi:10.1029/2006JD007515.

Tagaris, E., K. Manomaiphiboon, K.-J. Liao, L. R. Leung, J.-H. Woo, S. He, P. Amar, A. G. Russell, 2007: Impacts of Global Climate Change and Emissions on Regional Ozone and Fine Particulate Matter Concentrations over the United States, J. Geophys. Res., 112 (D14), D14312.

Tarasick, D. W. M. D. Moran, A. M. Thompson, T. Carey-Smith, Y. Rochon, V. S. Bouchet, W. Gong, P. A. Makar, C. Stroud, S. Ménard, L.-P. Crevier, S. Cousineau, J. A. Pudykiewicz, A. Kallaur, R. Moffet, R. Ménard, A. Robichaud, O. R. Cooper, S. J. Oltmans, J. C. Witte, G. Forbes, B. J. Johnson, J. Merrill, J. L. Moody, G. Morris, M. J. Newchurch, F. J. Schmidlin, E. Joseph, 2007: Comparison of Canadian air quality forecast models with tropospheric ozone profile measurements above midlatitude North America during the IONS/ICARTT campaign: Evidence for stratospheric input, J. Geophys. Res., 112, D12S22, doi:10.1029/2006JD007782

Tong, D.Q. and D.L. Mauzerall, 2006: Spatial Variability of Summertime Tropospheric Ozone over the Continental United States: Implications of an evaluation of the CMAQ model , Atmospheric Environment, 40, 3041–3056.

UCAR, 2007a: The NCAR Community Climate Model (CCM3). Available at http://www.cgd.ucar.edu/cms/ccm3/.

UCAR, 2007b: Community Climate Sytem Model, The Community Atmosphere Model. Available at http://www.ccsm.ucar.edu/ models/atm-cam.

UCAR, 2007c: Global Chemical Transport Modeling, MOZART-3. Available at http://gctm.acd.ucar.edu/mozart/models/m3/ index.shtml.

UCAR, 2007d: Atmospheric Chemistry Division, Community Data. Available at http://www.acd.ucar.edu/Data/.

UCAR, 2007e: Earth Systems Modeling Framework. Available at http://www.esmf.ucar.edu/.

Yu, S.C., R. Mathur, D. Kang, K. Schere, J. Pleim, and T.L. Otte, 2007: A detailed evaluation of the Eta-CMAQ forecast model performance for O3, its related precursors, and meteorological parameters during the 2004 ICARTT study, J. Geophys. Res., 112, D12S14, doi:10.1029/2006JD007715.

Zhang, F., N. Bei, J. W. Nielsen-Gammon, G. Li, R. Zhang, A. Stuart, and A. Aksoy, 2007: Impacts of meteorological uncertainties on ozone pollution predictability estimated through meteorological and photochemical ensemble forecasts, J. Geophys. Res., 112, D04304, doi:10.1029/2006JD007429.

Zhang, Y., P. Liu, B. Pun, C. Seigneur, 2006: A comprehensive performance evaluation of MM5-CMAQ for the summer 1999 southern oxidants study episode, Part I. Evaluation protocols, databases and meteorological predictions. Atmospheric Environment, 40, 4839–4855, doi:10.1016/j.atmosenv. 2005.12.043.

Chapter 3 References – Decision Support for Assessing Hybrid Renewable Energy Systems

Elliott, D. L., C. G. Holladay, W. R. Barchet, H. P. Foote, and W. R. Sandusky, 1987. Wind Energy (MERRA – http://gmao.gsfc.nasa.gov/research/merra/intro.php)Washington. DOE/CH 10093-4, March, 1987.

Solar and Wind Energy Resource Assessment (SWERA) Position Paper, 2003: Methodological Issues Related to Aerosol Data. Prepared by Dr. Christian Gueymard for the National Renewable Energy Laboratory.

Hansen, M.C., R. S. DeFries, J. R. G. Townshend, M. Carroll, C. Dimiceli, and R. A. Sohlberg, 2003. Global Percent Tree Cover at a Spatial Resolution of 500 Meters: First Results of the MODIS Vegetation Continuous Fields Algorithm. Earth Interactions 7(10):1–15.

Jennings, Michael and J. Michael Scott, 1997: Official Description of the GAP Analysis Program. http://gapanalysis.nbii.gov/portal/server.pt/gateway/PTARGS_0_2_1021_200_458_43/http%3B/gapcontent1%3B7087/publishedcontent/publish/public_sections/gap_home_sections/descriptionofficial/highlights_content.html.

Kalnay, E. M. Kanamitsu, R. Kistler, W. Collins, D. Deaven, L. Gandin, M. Iredell, S. Saha, G. White, J. Woollen, Y. Zhu, M. Chelliah, W. Ebisuzaki, W. Higgins, J. Janowiak, K. C. Mo, C. Ropelewski, J. Wang, A. Leetmaa, R. Reynolds, Roy Jenne, and Dennis Joseph,1996: The NMC/NCAR reanalysis project, Bull. Am. Meteor. Soc., 77, 437–471.

Koepke, P., M. Hess, I. Schult, and E. P. Shettle, 1997. Global Aerosol Data Set. Report No. 243, Max- Planck-Institut fur Meteorologie, Hamburg, ISSN 0937-1060.

Lambert, Tom, Paul Gilman, Peter Lilienthal., 2006. Micropower System Modeling with HOMER. In Felix A Farret, M Godoy Simoes. Integration of Alternative Sources of Energy. John Wiley and Sons, Inc. Hoboken, New Jersey. 379-416.

NASA Global Modeling and Assimilation Office, 2007: MERRA: Modern-Era Retrospective Analysis for Research and Applications. Available at http://gmao.gsfc.nasa.gov/research/merra/intro.php.

NOAA Earth Systems Research Laboratory, Physical Sciences Division 2007: The NCEP/NCAR Reanalysis Project at the NOAA/ESRL Physical Science Division. Available at http://www.cdc.noaa.gov/cdc/reanalysis/.

National Center for Environmental Prediction, 2007: North American Regional Reanalysis (NARR) Homepage, 2007. Available at http://www.emc.ncep.noaa.gov/mmb/rreanl/.

Onogi, Kazutoshi, Hiroshi Koide, Masami Sakamoto, Shinya Kobayashi, Junichi Tsutsui, Hiroaki Hatsushika, Takanori Matsumoto, Nobuo Yamazaki, Hirotaka Kamahori, Kiyotoshi Takahashi, Koji Kato, Ryo Oyama, Tomoaki Ose, Shinji Kadokura, Koji Wada, 2005. JRA-25: Japanese 25-year re-analysis project - progress and status. Quart. J. R. Meteorol. Soc., 131, 2961–3012 131(613):3259–3268.

Perez, R., P. Ineichen, K. Moore, M. Kmiecik, C. Chain, R. George, and F. Vignola, 2002: A New Operational Satellite-to-Irradiance Model. *Solar Energy* 73(5), pp. 307–317.

Renné, David S., Richard Perez, Antoine Zelenka, Charles Whitlock, and Roberta DiPasquale, 1999: Use of Weather and Climate Research Satellites for Estimating Solar Resources. Chapter 5 in Advances in Solar Energy, Volume 13, Edited by D. Yogi Goswami and Karl W. Boer. The American Solar Energy Society, 2400 Central Ave. Suite G1, Boulder, Colorado 80301. Pp. 171–240.

Schwartz, M., R. George, and D. Elliott, 1999. The Use of Reanalysis Data for Wind Resource Assessment at the National Renewable Energy Laboratory. Proceedings, European Wind Energy Conference, Nice, France, March 1–5, 1999.

Uppala, S.M., Kållberg, P.W., Simmons, A.J., Andrae, U., da Costa Bechtold, V., Fiorino, M., Gibson, J.K., Haseler, J., Hernandez, A., Kelly, G.A., Li, X., Onogi, K., Saarinen, S., Sokka, N., Allan, R.P., Andersson, E., Arpe, K., Balmaseda, M.A., Beljaars, A.C.M., van de Berg, L., Bidlot, J., Bormann, N., Caires, S., Chevallier, F., Dethof, A., Dragosavac, M., Fisher, M., Fuentes, M., Hagemann, S., Hólm, E., Hoskins, B.J., Isaksen, L., Janssen, P.A.E.M., Jenne, R., McNally, A.P., Mahfouf, J.-F., Morcrette, J.-J., Rayner, N.A., Saunders, R.W., Simon, P., Sterl, A., Trenberth, K.E., Untch, A., Vasiljevic, D., Viterbo, P., and Woollen, J. 2005: The ERA-40 re-analysis. Quart. J. R. Meteorol. Soc., 131, 2961-3012. Available at http://www.ecmwf.int/research/era/.

Chapter 4 References – Decision Support for Public Health

Beck, L.R. M.H Rodriguez, S.W. Dister, A.D. Rodriguez, R.K. Washino, D.R. Roberts and M.A. Spanner 1997: Assessment of a remote sensing-based model for predicting malaria transmission risk in villages of Chiapas, Mexico. *American Journal of Tropical Medicine and Hygiene* 56: 99–107.

Brownstein, J.S., T.R. Holford and D. Fish. 2003: A climate-based model predicts the spatial distribution of Lyme disease vector Ixodes scapularis in the United States. *Environmental Health Perspectives* 111: 1152–1157.

Brownstein, J.S., T.R. Holford and D. Fish 2005a: Effect of climate change on Lyme disease risk in North America. *EcoHealth* 2:38–46.

Brownstein, J.S., D. K Skelly, T.R. Holford and D. Fish. 2005b: Forest fragmentation predicts local scale heterogeneity of Lyme disease risk. *Oecologia* 146: 469–475

Fox,D. 2007: Back to the no-analog future? *Science* 316:823–825

Glass, G.E. 2007: Rainy with a chance of plague: forecasting disease outbreaks from satellites. *Future Virology* 2:225–229

Gubler, D.J. 2004: The changing epidemiology of yellow fever and dengue 1900 to 2003: full circle? Comparative Immunology Microbiology and Infectious Diseases 27:319–330.

Gubler, D.J., P. Reiter, K.L. Ebi, W. Yap, R. Nasci and J.A. Patz 2001: Climate variability and change in the United States: potential impacts on vector- and rodent-borne diseases. *Environmental Health Perspectives* 109:223.

Huntingford, C., D. Hemming, J.H.C. Gash, N Gedney, and P.A. Nuttall (2007) Impact of climate change on health: what is required of climate modellers? Trans. Royal Soc. Trop. Med. Hyg., 101, 97–103.

Lilienfeld, AM & DE Lilienfeld. 1980. Foundations of Epidemiology 2nd Edition. Oxford University Press, New York, 375 pp.

Linthicum, K.J., C.L. Bailey, F.G. Davies, and C.J. Tucker 1987: Detection of Rift Valley fever viral activity in Kenya by satellite remote sensing imagery. *Science* 235:1656–1659.

Malouin, R, P Winch, E Leontsini, G Glass, D Simon, EB Hayes & BS Schwartz. 2003. Longitudinal evaluation of an educational intervention to prevent tick bites in an area of endemic Lyme disease in Baltimore County, Maryland. Am J Epidemiol 157:1039–1051.

Piesman, J. and L. Gern 2004: Lyme borreliosis in Europe and North America. *Parasitology* 129:191–220.

Selvin, S. 1991: Statistical Analysis of Epidemiologic Data. Oxford University Press, New York, 375 pp.

Chapter 5 References – Water Management

Andreadis, K. and D. Lettenmaier, 2006: Trends in 20th century drought over the continental United States. *Geophysical Research Letters* 33, L10403.

Angel, J.R. and F.A. Huff, 1995: Seasonal distribution of heavy rainfall events in the Midwest. *Journal of Water Resources Planning and Management* 121, 110–115.

Baldwin, C. and U. Lall, 1999: Seasonality of streamflow: the upper Mississippi River, *Water Resources Research* 35(4), 1143.

Barnett, T., R. Malone, W. Pennell, D. Astammer, B. Demter, and W. Washington, 2004: The effects of climate change on water resources in the West: introduction and overview. *Climatic Change* 62, 1–11.

Beard, D., 1993: *Blueprint for Reform: The Commissioner's Plan for Reinventing Reclamation.* Bureau of Reclamation, Washington, D.C.

Benestad, R.E., 2004: Empirical-statistical downscaling in climate modeling. *EOS, Transactions, American Geophysical Union* 85, 417–422.*

Boroughs, C.B. and E.A. Zagona. 2002: Daily flow routing with the Muskingum-Cunge method in the Pecos River RiverWare Model, *Proceedings of the Second Federal Interagency Hydrologic Modeling Conference,* Las Vegas, NV.

Bradley, A., S. Schwartz, and T. Hashino, 2004: Distributions-oriented verification of ensemble streamflow predictions, Journal of Hydrometeorology 5(3), 532-545.

Bureau of Reclamation, 1992: *A Long Term Framework for Water Resource Management, Development, and Protection.* U.S. Department of Interior, Washington, DC.

Carroll, T., 1985: Snow surveying, in *Yearbook of Science and Technology,* pp. 386-388, McGraw-Hill, New York, N.Y.

Carroll, T., 1999: personal communication, National Operational Hydrologic Remote Sensing Center, National Weather Service.*

Carron, J., E. Zagona, and T. Fulp, 2006: Modeling uncertainty in an object-oriented reservoir operations model. J. *Irrig. and Drain. Engrg.,* 132(2), 104–111.

Changnon, S.A. (1990) The dilemma of climatic and hydrologic forecasting for the Great Lakes. In: *Proceedings of The Great Lakes Water Level Forecast and Statistics Symposium,* H.C. Hartmann and M.J. Donahue (Eds.), Great Lakes Commission, Ann Arbor, MI, pp. 13–25.

Changnon, D., 2000: Who used and benefited from the El Nino forecasts? In: *El Nino 1997-1998: The Climate Event of the Century,* S.A. Changnon (Ed.), Oxford University Press, New York, NY, pp. 109–135.

Christensen, N. and D.P. Lettenmaier, 2006: A multimodel ensemble approach to assessment of climate change impacts on the hydrology and water resources of the Colorado River basin, *Hydrology and Earth System Sciences,* 3, 1–44.

Christensen, N.S., A.W. Wood, N. Voisin, D.P. Lettenmaier, and R.N. Palmer, 2004: Effects of climate change on the hydrology and water resources of the Colorado River Basin. *Climatic Change* 62, 337–363.

Clark, M.P., L.E. Hay, G.J. McCabe, G.H. Leavesley, and R.L. Wilby, 1999: Towards the use of atmospheric forecasts in hydrologic models, I, Forecast drift and scale dependencies. *EOS Transactions AGU 80* Fall Meeting Supplement, Abstract H32G-10, F406-407.*

Congressional Budget Office, 1997: *Water Use Conflicts in the West: Implications of Reforming the Bureau of Reclamation's Water Supply Policies,* Congressional Budget Office, Washington, DC.

Croley, T.E., 1990: Laurentian Great Lakes double-CO2 climate change hydrological impacts. *Climatic Change* 17, 27–48.

Croley, T., F. Quinn, K. Kunkel, and S. Changnon, 1998: Great Lakes hydrology under a transposed climate. *Climatic Change* 38, 405–433.

Davis, R. E. and T. Pangburn, 1999: Development of new snow products for operational water control and management in the Kings River Basin, California. *EOS Transactions AGU 81*, Spring Meeting Supplement, Abstract H22D-07, S110.*

Douglas, E.M., R.M. Vogel, and C.N. Kroll, 2000: Trends in flood and low flows across the *U.S. Journal of Hydrology* 240, 90–105.

Duan, Q., H. V. Gupta, S. Sorooshian, A. N. Rousseau, and R. Turcotte, (eds.) 2002: *Calibration of Watershed Models*, American Geophysical Union, Washington, D. C.

Endreny, T., B. Felzer, J.W. Shuttleworth, and M. Bonell, 2003: Policy to coordinate watershed hydrological, social, and ecological needs: the HELP Initiative. In: *Water: Science, Policy, and Management*, R. Lawford, D. Fort, H. Hartmann, and S. Eden (Eds.), American Geophysical Union, Washington, DC, pp. 395–411.

Environmental Protection Agency, 1989: *The Potential Effects of Global Climate Change on the United States.* Report to Congress. J.B. Smith and D. Tirpak, (Eds), EPA Office of Policy, Planning and Evaluation, Washington, D.C.

Eschenbach, E.A., T. Magee, E. Zagona, M. Goranflo, and R. Shane, 2001: Goal Programming Decision Support System for Multiobjective Operation of Reservoir Systems. *Journal of Water Resources Planning and Management*, 127, 71–141.

Ezurkwal, B., 2005: The role and importance of paleohydrology in the study of climate change and variability. In: *Encyclopedia of Hydrological Sciences*, M.G. Anderson (Ed.), John Wiley and Sons, Ltd., West Sussex, UK.

Franz, K., H.C. Hartmann, S. Sorooshian, and R. Bales, 2003: An evaluation of National Weather Service ensemble streamflow predictions for water supply forecasting in the Colorado River Basin. *Journal of Hydrometeorology* 4, 1105–1118.

Frederick, K., D. Major, and E. Stakhiv, (Eds.) 1997: *Climate Change and Water Resources Planning Criteria.* Kluwer Academic Publishers, Dordrecht, Netherlands.

Frevert, D.K., M.S. Cowan, and W.L. Lane, 1989: Use of stochastic hydrology in reservoir operation. *Journal of Irrigation and Drainage Engineering* 115, 334–343.

Frevert, D., T. Fulp, E. Zagona, G. Leavesley, and H. Lins, 2006: *Watershed and River Systems Management Program: Overview of Capabilities.* J. Irrig. and Drain. Engrg. 132(2), 92–97.

Gamble, J.L., J. Furlow, A.K. Snover, A.F. Hamlet, B.J. Morehouse, H. Hartmann, and T. Pagano, 2003: Assessing the impact of climate variability and change on regional water resources: the implications for stakeholders. In: *Water: Science, Policy, and Management*, R. Lawford, D. Fort, H. Hartmann, and S. Eden (Eds.), American Geophysical Union, Washington, DC, pp. 341–368.

Garen, D.C., 1992: Improved techniques in regression-based streamflow volume forecasting, *J. Water Resour. Planning and Manag.*, 118, 654–670.

Georgakako, A., 2006: Decision-Support Systems for integrated water resources management with an application to the Nile Basin. In: *Topics on System Analysis and Integrated Water Resources Management*, A. Castelletti and R. Soncini-Sessa (Eds.), Elsevier, New York, NY.

Georgakakos, A., H. Yao, M. Mullusky, and K. Georgakakos, 1998: Impacts of climate variability on the operational forecast and management of the Upper Des Moines River Basin, Water Resour. Res., 34, 799–821.

Georgakakos, K., E. Shamir, S. Taylor, T. Carpenter, and N. Graham, 2004: Integrated Forecast and Reservoir Management INFORM - A Demonstration for Northern California Phase 1 Progress Report. HRC Limited Distribution Rept. No. 17, Hydrologic Research Center, San Diego, CA. *

Georgakako, K., N. Graham, T. Carpenter, A. Georgakakos, and H. Yao, 2005: Integrating climate-hydrology forecasts and multi-objective reservoir management for northern California. EOS Transactions 86, 122, 127.*

Gilmore, A., T. Magee, T. Fulp, and K. Strezepek, 2000: Multiobjective optimization of the Colorado River. Proceedings of the ASCE 2000 Joint Conference on Water Resources Engineering and Water Rousources Planning and Management, Minneapolis, MN.*

Glantz, M.H., 1982: Consequences and responsibilities in drought forecasting- the case of Yakima, 1977, Water Resour. Res., 18, 3–13.

Gleick, P.H. and D.B. Adams, 2000: Water: The Potential Consequences of Climate Variability and Change for Water Resources of the United States. Pacific Institute, Oakland, CA.

Grantz, K., B. Rajagopalan, E. Zagona, and M. Clark, 2007: Water management applications of climate-based hydrologic forecasts: case study of the Truckee-Carson River basin, Nevada. Journal of Water Resources Planning and Management.

Grayson, R., and G. Bloschl, 2000: Spatial Patterns in Catchment Hydrology: Observations and Modelling, Cambridge University Press, Cambridge, U. K.

Hamlet, A. F., and D. P. Lettenmaier, 1999: Columbia River streamflow forecasting based on ENSO and PDO climate signals, J. Water Resour. Planning and Manag., 125, 333–341.

Hansen, J., M. Sato, R. Ruedy, A. Lacis, K. Asamoah, K. Beckford, S. Borenstein, E. Brown, B. Cairns, B. Carlson, B. Curran, S. de Castro, L. Druyan, P. Etwarrow, T. Ferede, M. Fox, D. Gaffen, J. Glascoe, H. Gordon, S. Hollandsworth, X. Jiang, C. Johnson, N. Lawrence, J. Lean, J. Lerner, K. Lo, J. Logan, A. Luckett, M. P. McCormick, R. McPeters, R. Miller, P. Minnis, I. Ramberran, G. Russell, P. Russell, P. Stone, I. Tegen, S. Thomas, L. Thomason, A. Thompson, J. Wilder, R. Willson, and J. Zawodny, 1997: Forcings and chaos in interannual to decadal climate change, J. Geophy. Res., 102, 25679-25720.

Harding, B.J., T.B. Sangoyomi, and E.A. Payton, 1995: Impacts of severe sustained drought on Colorado River water resources. Water Resources Bulletin 31, 815–824.

Hartmann, H.C., 1990: Impacts on Laurentian Great Lakes levels. Climatic Change 17, 49–68.

Hartmann, H.C., T.C. Pagano, S. Sorooshian, and R. Bales, (2002a): Confidence builders: evaluating seasonal climate forecasts from user perspectives. Bulletin of the American Meteorological Society 83(5), 683–698.

Hartmann, H.C., R. Bales, and S. Sorooshian, (2002b): Weather, climate, and hydrologic forecasting for the U.S. Southwest: a survey. Climate Research 21, 239–258.

Hartmann, H., A. Bradley, and A. Hamlet, (2003): Advanced hydrologic prediction for improving water management. In: Lawford, R., Fort, D., Hartmann, H., and S. Eden (Editors), Water: Science, Policy, and Management. Water Resources Monograph 16, American Geophysical Union, Washington, DC, pp.285–307.

Hartmann, H.C., 2005: Use of climate information in water resources management. In: Encyclopedia of Hydrological Sciences, M.G. Anderson (Ed.), John Wiley and Sons Ltd., West Sussex, UK, Chapter 202.

Hoerling, M. and J. Eischeid, 2007: Past peak water in the Southwest, Southwest Hydrology 6(10),18–19, 35.*

Hydrological Sciences Branch, 2007: Evaluation Report for AWARDS ET Toolbox and RiverWare Decision-support Tools. NASA Goddard Space Flight Center, Greenbelt, MD, 28 pp. (URL: http://wmp.gsfc.nasa.gov/projects/project_RiverWare.php)*

IPCC, 1990: Scientific Assessment of Climate Change: Report of Working Group I to the First Assessment Report of the IPCC. Cambridge University Press, Cambridge.

IPCC, 1995a: Climate Change 1995: IPCC Second Assessment. Cambridge University Press, Cambridge.

IPCC, 1995b: Impacts, Adaptations and Mitigations: Contributions of Working Group II to the Second Assessment Report of the IPCC. Cambridge University Press, Cambridge.

IPCC, 2001a: Climate Change 2001: Synthesis Report. Third Assessment Report of the IPCC. Cambridge University Press, Cambridge.

IPCC, 2001b: Impacts, Adaptations, and Vulnerability: Contribution of Working Group II to the Third Assessment Report of the IPCC. Cambridge University Press, Cambridge.

IPCC, 2007: Climate Change 2007: Climate Change Impacts, Adaptation and Vulnerability. Working Group II Contribution to the IPCC Fourth Assessment Report (http://www.gtp89.dial.pipex.com/chpt.htm).

Jacobs, K., 2002: Connecting Science, Policy, and Decision-Making: A Handbook for Researchers and Science Agencies. Office of Global Programs, National Oceanic and Atmospheric Administration, Silver Spring, MD.

Jacobs, K. and R. Pulwarty, 2003: Water resource management: science, planning and decision-making. In: Water: Science, Policy, and Management, R. Lawford, D. Fort, H. Hartmann, and S. Eden (Eds.), American Geophysical Union, Washington, DC, pp. 177–204.

Jerla, C., 2005: An Analysis of Coordinated Operation of Lakes Powell and Mead under Low Reservoir Conditions. M.S. Thesis, University of Colorado-Boulder, Boulder, CO, 187 pp.

Kenney, D., 1995: Institutional options for the Colorado River. Water Resources Bulletin 31(5), 837–850.

Kruger, A., S. Khandelwal, and A. Bradley, 2007: AHPSVER: A web-based system for hydrologic forecast verification. Computers and Geosciences 33(6), 739–748.

Lawford, R., R. Try, and S. Eden, 2005: International research programs in global hydroclimatology. In: Encyclopedia of Hydrological Sciences, M.G. Anderson (Ed.), John Wiley and Sons, Ltd., West Sussex, UK.

Lee, D.H., Quinn, F.H., D. Sparks, and J.C. Rassam, 1994: Modification of Great Lakes regulation plans for simulation of maximum Lake Ontario outflows. Journal of Great Lakes Research 20, 569–582

Lee, D.H., T.E. Croley, II, and F.H. Quinn, 1997: Lake Ontario regulation under transposed climates. Journal of the American Water Resources Association 33, 55–69

Lee, D.H., 1999: Institutional and technical barriers to implementing risk-based water resources management: a case study. Journal of Water Resources Planning and Management 125, 186–193.

Lettenmaier, D.P., 2003: The role of climate in water resources planning and management. In: Water: Science, Policy, and Management, R. Lawford, D. Fort, H. Hartmann, and S. Eden (Eds.), American Geophysical Union, Washington, DC, pp. 247–266.

Lettenmaier, D.P., E.F. Wood, and J.R. Wallis, 1994: Hydro-climatological trends in the continental United States, 1948-88. Journal of Climate 7, 586–607.

Lettenmaier, D., A. Wood, R. Palmer, E. Wood, and E. Stakhiv, 1999: Water resources implications of global warming: a U.S. regional perspective. Climatic Change 43, 537–579.

Lins, H.F. and J.R. Slack, 1999: Streamflow trends in the United States. Geophysical Research Letters 26, 227–230

Lins, H.F. and E.Z. Stakhiv, 1998: Managing the nation's water in a changing climate. Journal of the American Water Resources Association 34, 1255–1264.

Lofgren, B.M., F.H. Quinn, A.H. Clites, R.A. Assel, A.J. Eberhardt, and C.L. Luukkonen, 2002: Evaluation of potential impacts on Great Lakes water Rresources based on climate scenarios of two GCMs. Journal of Great Lakes Research. 28, 537–554.

Loomis, J., J. Koteen, and B. Hurd, 2003: Economic and institutional strategies for adapting to water-resource effects of climate change. In: Water and Climate in the Western United States. W. Lewis (Ed.), University Press of Colorado, Boulder, CO, pp. 235–249.

Mantua, N., S. Hare, Y. Zhang, J. M. Wallace, and R. Francis, 1997: A Pacific interdecadal climate oscillation with impacts on salmon production, Bull. Amer. Meteor. Soc., 78, 1069–1079.

Matalas, N.C., 1997: Stochastic hydrology in the context of climate change. Climatic Change 37, 89–101.

Maurer, E., 2007: Uncertainty in hydrologic impacts of climate change in the Sierra Nevada, California, under two emissions scenarios. Climatic Change 82, 309–325.

Miles, E. L., A.K. Snover, A.F. Hamlet, B. Callahan, and D. Fluharty, 2000: Pacific northwest regional assessment: the impacts of climate variability and change on the water resources of the Columbia river basin. Journal of the American Water Resources Association 36, 399–420.

Milly, P., K. Dunne, and A. Vecchia, 2005: Global patterns of trends in streamflow and water availability in a changing climate, Nature 438, 347–350.

NRC, 1995) Flood Risk Management and the American River Basin: An Evaluation. National Academy Press, Washington, DC.

NRC, 1998a: GCIP: A Review of Progress and Opportunities. National Academy Press, Washington, DC.

NRC, 1998b: Hydrologic Sciences: Taking Stock and Looking Ahead. National Academy Press, Washington, DC.

NRC, 1999a: Making Climate Forecasts Matter. National Academy Press, Washington, DC.

NRC, 1999b: A Vision for the National Weather Service: Road Map for the Future. National Academy Press, Washington, DC.

NRC, 1999c: Hydrologic Science Priorities for the U.S. Global Change Research Program: An Initial Assessment, National Academy Press, Washington, DC.

NRC, 2004: Analytical Methods and Approaches for Water Resources Project Planning, National Academies Press, Washington, DC, 151 pp.

NASA, 2005a: Earth-Sun System Applied Sciences Program Water Management Program Element FY2005–2009 Plan, Washington, DC.

NASA, 2005b: Water Management Annual Report, Goddard Space Flight Center, December.*

NASA, 2006: Applied sciences program, The Subcommittee on Hydrology Newsletter 1, 12–14.*

NASA, 2007: Water Management Progress Report Jan-Mar 2007, Goddard Space Flight Center, 27 April.*

National Assessment Synthesis Team (2000) Climate Change Impacts on the United States: The Potential Consequences of Climate Variability and Change. U.S. Global Change Research Program, Washington, DC.

National Hydrologic Warning Council (2002) Use and Benefits of the National Weather Service River and Flood Forecasts. National Weather Service Office of Hydrologic Development, Silver Spring, Md.

Neumann, D., E. Zagona, and B. Rajagopalan, 2006: A Decision-Support System to manage summer stream temperatures. Journal of the American Water Resources Association 42, 1275–1284.

Olsen, J.R., J.H. Lambert, and Y.Y. Haimes, 1998: Risk of extreme events under nonstationary conditions. Risk Analysis 18, 497–510.

Olsen, J.R., J.R. Stedinger, N.C. Matalas, and E.Z. Stakhiv, 1999: Climate variability and flood frequency estimation for the upper Mississippi and lower Missouri rivers. Journal of the American Water Resources Association 35, 1509–1523.

Office of Global Programs, 2004: Regional Integrated Sciences and Assessments. National Oceanic and Atmospheric Administration, http://www.risa.ogp.noaa.gov, 3 June 2004.*

Oudin, L., C. Perrin, T. Mathevet, V. Andreassian, and C. Michel, 2006: Impact of biased and randomly corrupted inputs on the efficiency and the parameters of watershed models. Journal of Hydrology pp. 1–2, 62–83.

Pagano, T.C., H.C. Hartmann, and S. Sorooshian, 2001: Using climate forecasts for water management: Arizona and the 1997-98 El Nino, Journal of the American Water Resources Association 37, 1139–1153.

Pagano, T.C., H.C. Hartmann, and S. Sorooshian, 2002: Use of climate forecasts for water management in Arizona: a case study of the 1997-98 El Niño. Climate Research 21, 59–269.

Pagano, T., D. Garen, and S. Sorooshian, 2004: Evaluation of official Western U.S. seasonal water supply outlooks, 1922-2002. Journal of Hydrometeorology 5(5), 896–909.

Payne, J.T., A.W. Wood, A.F. Hamlet, R.N. Palmer, and D.P. Lettenmaier, 2004: Mitigating the effects of climate change on the water resources of the Columbia River Basin. Climatic Change 62, 233–256.

Piechota, T.C., and J.A. Dracup, 1999: Long range streamflow forecasting using El Niño-Southern Oscillation indicators, J. Hydrol. Engineer., 4, 144–151.

Piechota, T.C., F.H.S. Chiew, J.A. Dracup, and T.A. McMahon, 2001: Development of an exceedance probability streamflow forecast using the El Niño-Southern Oscillation, J. Hydrol. Engineer., 4, 20–28.

Pielke, R.A., Jr., 1995: Usable information for policy: an appraisal of the U.S. global change research program. Policy Sciences 38, 39–77.

Pielke, R.A., Jr., 2001: The Development of the U.S. Global Change Research Program: 1987 to 1994. Policy Case Study, National Center for Atmospheric Research, Boulder, CO.*

Prairie, J.R., 2006: Stochastic nonparametric framework for basin wide streamflow and salinity modeling: application for the Colorado River basin. Civil Environmental and Architectural Engineering Ph.D. Dissertation, University of Colorado, Boulder, CO.

Pulwarty, R.S., 2002: Regional Integrated Sciences and Assessment Program. Office of Global Programs, National Oceanic and Atmospheric Administration, Silver Spring, MD.

Pulwarty, R.S. and K.T. Redmond, 1997: Climate and salmon restoration in the Columbia River basin: the role and usability of seasonal forecasts. Bulletin of the American Meteorological Society 78, 381–397.

Quinn, F.H., 1981: Secular changes in annual and seasonal Great Lakes precipitation, 1854–1979, and their implications for Great Lakes water resources studies. Water Resources Research 17, 1619–1624.

Quinn, F.H., 2002: Secular changes in Great Lakes water level changes. Journal of Great Lakes Research 28, 451–465.

Rayner, S., D. Lach, and H. Ingram, 2005: Weather forecasts are for wimps: why water resource managers do not use climate forecasts. Climatic Change 69, 197–227.

Reitsma, R.F., 1996: Structure and support of water resources management and decision making, Journal of Hydrology, 177(1), 253-268.

Reitsma, R., I. Zigurs, C. Lewis, V. Wilson, and A. Sloane, 1996: Experiment with simulation models in water resources negotiations, Journal of Water Resources Management and Planning, ASCE 122, 64–70.

Salas, J.D., 1993: Analysis and modeling of hydrologic time series. In: Handbook of Hydrology, D.R. Maidment (Ed.), McGraw-Hill, Inc., New York, NY, Chapter 19.

Saunders, J.F., III and W.M. Lewis, Jr., 2003: Implications of climatic variability for regulatory low flows in the South Platte River Basin, Colorado. Journal of the American Water Resources Association 39, 33–45.

Seager, R., M. Ting, I. Held, Y. Kushnir, J. Lu, G. Vecchi, H.P. Huang, N. Harnik, A. Leetma, N.C. Lau, C. Li, J. Velez, and N. Naik, 2007: Model projections of an imminent transition to a more arid climate in Southwestern North America, Science 316(5828), 1181–1184.

Schaake, J., T. Hamill, R. Buizza, and M. Clark, 2007: HEPEX, the Hydrological Ensemble Prediction Experiment. Bulletin of the American Meteorological Society, 88(10:1541–1547.

Smith, J. B., K.C. Hallet, J. Henderson, and K.M. Strzepek, 2007: Expanding the tool kit for water management in an uncertain climate. Southwest Hydrology, 6, 24–35, 36.*

Snover, A.K., A.F. Hamlet, and D.P. Lettenmaier, 2003: Climate change scenarios for water planning studies: pilot applications in the Pacific Northwest. Bulletin of the American Meteorological Society 84, 1513–1518.

Sousounis, P., G. Albercook, D. Allen, J. Andresen, A. Brooks, D. Brown, H.H. Cheng, M. Davis, J. Lehman, J. Lindeberg, J. Root, K. Kunkel, B. Lofgren, F. Quinn, J. Price, T.D. Stead, J. Winkler, and M. Wilson, 2000: Preparing for a Changing Climate: The Potential Consequences of Climate Variability and Change for the Great Lakes. U.S. Global Change Research Program, Washington, DC.

Stakhiv, E., 2003: What can water managers do about climate variability and change? In: Water and Climate in the Western United States. W. Lewis (Ed.), University Press of Colorado, Boulder, CO, pp. 131–142.

Tarboton, D., 1995: Hydrologic scenarios for severe sustained drought in the Southwestern United States. Water Resources Bulletin 31(5), 803-813.

Urbanas, B.R. and L.A. Roesner, 1993: Hydrologic design for urban drainage and flood control. In: Handbook of Hydrology, D.R. Maidment (Ed.), McGraw-Hill, Inc., New York, NY, Chapter 28.

U.S. Army Corps of Engineers, 1992: Guidelines for Risk and Uncertainty Analysis in Water Resources Planning, Volumes I and II. IWR Report 92-R-1, 92-R-2. Institute for Water Resources, Fort Belvoir, VA.

U.S. Department of Interior, 2007: Colorado River Interim Guidelines for Lower Basin Shortages and Coordinated Operations for Lake Powell and Lake Mead, Draft Environmental Impact Statement, Volume 1, Bureau of Reclamation, Boulder City, NV. (URL: http://www.usbr.gov/lc/region/programs/strategies/draftEIS/index.html)

VanRheenen, N., A.W. Wood, R.N. Palmer, and D.P. Lettenmaier, 2004: Potential implications of PCM climate change scenarios for Sacramento-San Joaquin River basin hydrology and water resources. Climatic Change 62, 257–281.

Vicuna, S., E. Maurer, B. Joyce, J. Dracup, and D. Purkey, 2007: The sensitivity of California water resources to climate change scenarios. Journal of the American Water Resources Association 43, 482–498.

Vorosmarty, C., D. Lettenmaier, C. Leveque, M. Meybeck, C. Pahl-Wostl, J. Alcamo, W. Cosgrove, H. Grassl, H. Hoff, P. Kabat, F. Lansigan, R. Lawford, R. Naiman, 2004: Humans transforming the global water system. EOS, Transactions, American Geophysical Union 85, 509–514.

Walker, A. E., and B. E. Goodison, 1993: Discrimination of wet snow cover using passive microwaver satellite data, Annals of Glaciology, 17, 307–311.

Ward, R.C., R. Pielke Sr., and J. Salas, (Eds.) 2003: Special Issue: Is Global Climate Change Research Relevant to Day-to-day Water Resources Managers? Water Resources Update 124.

Welles, E., S. Sorooshian, G. Carter, and B. Olsen, 2007: Hydrologic verification: a call for action and collaboration. Bulletin of the American Meteorological Society 88(4), 503–511.

Westrick, K. J. and C. F. Mass, 2001: An evaluation of a high resolution hydrometeorological modeling system for prediction of a cool-season flood event in a coastal mountainous watershed. Journal of Hydrometeorology 2, 161–180.

Westrick, K. J., P. Storck, and C. F. Mass, 2002: Description and evaluation of a hydrometeorologi¬cal forecast system for mountainous watersheds, Wea.and Forecasting, 17, 250–262.

Wheeler, K., T. Magee, T. Fulp, and E. Zagona, 2002: Alternative policies on the Colorado River. Proceedings of Natural Resources Law Center Allocating and Managing Water for a Sustainable Future: Lessons From Around the World, Natural Resources Law Center, University of Colorado, Boulder, CO.*

Wood, E., E.P. Maurer, A. Kumar, and D.P. Lettenmaier, 2002: Long-range experimental hydrologic forecasting for the eastern United States. Journal of Geophysical Research, Atmospheres 107, 4423–4429.

Wood, A.W., A. Hamlet, D.P. Lettenmaier, and A. Kumar, 2001: Experimental real-time seasonal hydro¬logic forecasting for the Columbia River Basin, Proc., 26th Annual Climate Diagnostics and Prediction Workshop, National Weather Service, PB92-167378, National Technical Information Service, Springfield, VA.

Woodhouse, C., and J.J. Lukas, 2006: Multi-Century Tree-Ring Reconstruction of Colorado Streamflow for water resource planning. Climatic Change, (78), 293-315.

Young, R.A., (Ed.) 1995: Special Issue: Managing the Colorado River in a severe sustained drought. Water Resources Bulletin 35, 779-944.

Zagona, E.A., T.J. Fulp, H. Goranflo, and R. Shane, 1998: RiverWare: a general river and reservoir modeling environment. Proceedings of the First Federal Interagency Hydrologic Modeling Conference, Las Vegas, NV, 19-23 April, 5,113–120. *

Zagona, E., T.J. Fulp, R. Shane, T. Magee, and H. Morgan Goranflo, 2001: RiverWare: a generalized tool for complex reservoir systems modeling. Journal of the American Water Resources Association 37, 913–929.

Zagona, E., T. Magee, D. Frevert, T. Fulp, M. Goranflo, and J. Cotter, 2005: RiverWare. In: Watershed Models, V. Singh & D. Frevert (Eds.), Taylor & Francis/CRC Press: Boca Raton, FL.

List of Figures and Tables by Chapter

Glossary, Acronyms, Symbols, and Abbreviations

3-D	Three-dimensional
AERONET	Aerosol RObotic NETwork
Ag	Agricultural
AgRISTARS	Agriculture and Resources Inventory Surveys through Aerospace Remote Sensing
AGRMET	Agricultural Meteorological Model
AOD	Aerosol Optical Depth
AOT	Aerosol Optical Thickness
AQS	Air Quality System
ASOS	Automated Surface Observation Stations
ASTER	Advanced Space-borne Thermal Emission and Reflection Radiometer
AVHRR	Advanced Very High Resolution Radiometer
AWARDS	Agricultural Water Resources Decision Support
BELD3	Biogenic Emissions Land Use Database Version 3
CADRE	Crop Condition Data Retrieval and Evaluation System
CADSWES	Center for Advanced Decision Support for Water and Environmental Systems
CALIPSO	Cloud-Aerosol Lidar and Infrared Pathfinder Satellite Observation
CCM	Community Climate Model
CCM3	NCAR Community Climate Model
CCSP	Climate Change Science Program
CDC	Disease Control and Prevention
CENR	Committee on Environment and Natural Resources Research
CFC	Chlorofluorocarbons
CM2	Climate Model 2
CMAQ	Community Multiscale Air Quality
CMAS	Community Modeling and Analysis System
COTS	Commercial Off-the-Shelf
DEM	Digital Elevation Models
DLR	German Aerospace Center (DLR) (German: Deutsches Zentrum für Luft- und Raumfahrt e.V.)
DSS	Decision-Support System
DSSPL	Decision-Support System to Prevent Lyme Disease
DST	Decision Support Tool
ENSO	El Nino-Southern Oscillation
EOS	Earth Observing System
EPA	Environmental Protection Agency
EROS	Earth Resources Observation Systems
ESRI	Environmental Science and Research Institute
ESRL	Earth Systems Research Laboratory
ESSL	Earth and Sun Systems Laboratory
ET	Evapotranspiration
FAS	Foreign Agricultural Service
GACP	Global Aerosol Climatology Project
GADS	Global Aerosol Dataset
GCM	Global Climate Model
GCTM	Global Chemistry Transport Models
GEO	Group on Earth Observations
GEOS	Goddard Earth Observing System
GEOS-Chem	Goddard Earth Observing System-Chemistry
GEOSS	Global Earth Observations System of Systems
GFDL	Geophysical Fluid Dynamics Laboratory
GhTOC	Hourly Total Ozone Column

GIMMS	Global Inventory Modeling and Mapping Studies
GIS	Geographic Information System
GISS	Goddard Institute for Space Studies
GMAO	Global Modeling and Assimilation Office
GOCART	Global Ozone Chemistry Aerosol Transport
GOES	Geostationary Environmental Operational Satellite
GPM	Global Precipitation Mission
GRACE	Gravity Recovery and Climate Experiment
GsT	Geospatial Toolkit
GTCM	Global Tropospheric Chemistry Model
GUI	Graphical User Interface
HIV	Human Immunodeficiency Virus
HOMER	Hybrid Optimization Model for Electric Renewables
HSPF	Hydrological Simulation Program – Fortran
Int'l	International
IPCC	Intergovernmental Panel on Climate Change
ISH	Integrated Surface Hourly
KAMM	Karlsruhe Atmospheric Mesoscale Model
km	Kilometer
LACIE	Large Area Crop Inventory Experiment
Landsat	Land Remote-Sensing Satellite
m	Meter
MISR	Multi-Angle Imaging Spectroradiometer
MM5	Mesoscale Model Version 5
MMS	Modular Modeling System
MODIS	Moderate Resolution Imaging Spectroradiometer
MOZART	Model of Ozone and Related Chemical Tracers
NASA	National Aeronautics and Space Administration
NCAR	National Center for Atmospheric Research
NCDC	National Climatic Data Center
NCEP	National Centers for Environmental Prediction
NEP	Net Ecosystem Productivity
NOAA	National Oceanic and Atmospheric Administration
NPOESS	National Polar-Orbiting Operational Environmental Satellite
NRC	National Research Council
NREL	National Renewable Energy Laboratory
NSTC	National Science and Technology Council
OCO	Orbiting Carbon Observatory
OMI	Ozone Monitoring Instrument
PAN	Peroxyacetyl Nitrate
PECAD	Production Estimate and Crop Assessment Division
PNNL	Pacific Northwest National Laboratory
PRMS	Precipitation Runoff Modeling System
Quickscat	Quick Scatterometer
RCM	Regional Climate Model
SAP	Synthesis and Assessment Product
SARS	Severe Acute Respiratory Syndrome
SLEUTH	Slope, Land Cover, Exclusions, Urban Areas, Transportation, Hydrologic
SPOT	Systeme Pour L'Observation de la Terre
SPOT-VEG	Systeme Pour L'Observation de la Terre-Vegetation
SRES	Special Report on Emissions Scenarios
SRTM	Shuttle Radar Topology Mission
SSE	Surface Meteorology and Solar Energy
SSE	Surface Meteorology and Solar Energy
SSM/I	Special Sensor Microwave/Imager

SSMR	Scanning Multichannel Microwave Radiometer
STAR	Science to Achieve Results
STEM	Sulfur Transport Eulerian Model
SWAT	Soil and Water Assessment Tool
SWERA	Solar and Wind Energy Resource Assessment
TOC	Total Ozone Content
TOMS	Total Ozone Mapping Spectrometer
TRMM	Tropical Rainfall Mapping Mission
U.S.	United States
UIUC	Unknown Sent email to Daewon
USDA	Department of Agriculture
USGS	United States Geological Survey
VIC	Variable Infiltration Capacity
WAsP	Wind Atlas Analysis and Application Program
WRAMS	Wind Resource Assessment Mapping System
WRDC	World Radiation Data Centre
WRF	Weather Research and Forecasting
WRF-Chem	Weather Research and Forecasting-Chemistry

Contact Information

Global Change Research Information Office
c/o Climate Change Science Program Office
1717 Pennsylvania Avenue, NW
Suite 250
Washington, DC 20006
202-223-6262 (voice)
202-223-3065 (fax)

The Climate Change Science Program incorporates the U.S. Global Change Research Program and the Climate Change Research Initiative.

To obtain a copy of this document, place an order at the Global Change Research Information Office (GCRIO) web site: http://www.gcrio.org/orders.

Climate Change Science Program and the Subcommittee on Global Change Research

William Brennan, Chair
Department of Commerce
National Oceanic and Atmospheric Administration
Director, Climate Change Science Program

Jack Kaye, Vice Chair
National Aeronautics and Space Administration

Allen Dearry
Department of Health and Human Services

Jerry Elwood
Department of Energy

Mary Glackin
National Oceanic and Atmospheric Administration

Patricia Gruber
Department of Defense

William Hohenstein
Department of Agriculture

Linda Lawson
Department of Transportation

Mark Myers
U.S. Geological Survey

Timothy Killeen
National Science Foundation

Patrick Neale
Smithsonian Institution

Jacqueline Schafer
U.S. Agency for International Development

Joel Scheraga
Environmental Protection Agency

Harlan Watson
Department of State

EXECUTIVE OFFICE AND OTHER LIAISONS

Stephen Eule
Department of Energy
Director, Climate Change Technology Program

Katharine Gebbie
National Institute of Standards & Technology

Stuart Levenbach
Office of Management and Budget

Margaret McCalla
Office of the Federal Coordinator for Meteorology

Rob Rainey
Council on Environmental Quality

Daniel Walker
Office of Science and Technology Policy

www.ingramcontent.com/pod-product-compliance
Lightning Source LLC
Chambersburg PA
CBHW081555170526
45166CB00009B/2713